Monograph Series of the Socio-Economic History Society, Japan

This monograph series is published by the Socio-Economic History Society, Japan, and Springer.

The aim of the series is to make works by Japanese scholars accessible to a wider readership and thereby enhance the global knowledge of Japanese and Asian scholarship in the fields of economic, social, and business history.

This book series will present English translations of outstanding recent academic research works, survey articles, and book reviews carefully selected from the Society's quarterly Japanese-language journal *Socio-Economic History*, which is the preeminent journal in its field in Japan, as well as from other publications of the Society. These will be edited under specific themes such as energy and the environment, the consumer society and the company system in postwar Japan, and the economic history of Japan in the early twentieth century. The content will include chapters on economic, social, and business history ranging geographically from Japan and Asia as a whole to Europe and the United States, and a small number of book reviews of recent academic works published in Japanese and other languages.

The Society was founded in 1930 and currently comprises more than 1,400 registered members, mainly academics, researchers and postgraduate students affiliated with universities and research institutions in Japan.

More information about this series at http://www.springer.com/series/13569

S. Sugiyama
Editor

Economic History of Energy and Environment

 Springer

Editor
S. Sugiyama
Professor Emeritus
Keio University
Japan

Former President
Socio-Economic History Society
Japan

ISSN 2364-2394   ISSN 2364-2408   (electronic)
Monograph Series of the Socio-Economic History Society, Japan
ISBN 978-4-431-55506-3   ISBN 978-4-431-55507-0   (eBook)
DOI 10.1007/978-4-431-55507-0

Library of Congress Control Number: 2015939799

Springer Tokyo Heidelberg New York Dordrecht London
© Socio-Economic History Society, Japan 2015

Printed on acid-free paper

Springer Japan KK is part of Springer Science+Business Media (www.springer.com)

Preface

This book is the first volume in a monograph series published by the Socio-Economic History Society, Japan, and Springer. It contains four recent articles on topics related to the history of energy and the environment in Japan, China, and Britain, and four short book reviews on recent academic works published in Japanese and English. Since they are English language versions of material originally published in Japanese, some revisions have been made to render them accessible to this new readership.

Environmental history is a major theme in global history. The relationship between economic growth, ecological destruction, and environmental pollution is clear. For example, timber played a vital role in industrial development as an energy resource for iron, pottery, and glass manufacturing, and as an industrial material for railway sleepers, pit props, and telegraphic and electric poles, as well for construction. However, the rising demand for timber led to deforestation on a global level, producing natural disasters such as flooding, and a drastic reduction in biodiversity.

The first two articles in this volume approach different aspects of the demand for timber in modern Japan. Chapter 1, by Sugiyama and Yamada, examines silk reeling, a traditional industry that was Japan's most important export industry until 1929. They analyse the relationship between deforestation and the development of the silk reeling industry in a district of Nagano Prefecture (central Japan) from the 1870s to the 1900s. The increasing energy demands of the industry caused the deforestation of common lands, producing a shortage of firewood. Reforestation was unsuccessful and it became necessary to transport firewood from neighbouring districts. Over time, the improvement in quality of steam boilers and the decline in the relative prices of coal relative to firewood facilitated a shift from firewood to coal.

By contrast, Chap. 2, by Yamaguchi, looks at timber in relation to the development of railways, a technology newly introduced from the West. Little research has been done on the use of timber as an industrial material. Demand for railway sleepers increased along with the formation of railway networks all over the world. Yamaguchi's case study of the supply of timber for use as rail sleepers in the

Japanese national railway network during the prewar period is therefore a unique contribution to the field of environmental history. In Meiji, Japan, railways were first introduced as a symbol of Western culture and came to play a crucial role alongside cotton spinning as one of the leading sectors in early industrialization. Japan's total railway mileage increased from 184 km in 1880 to 6,200 km in 1900. Budgetary reasons hindered government development of railways, leaving private railways to take the leading role in the railway boom of the 1880s onwards. After the nationalization of major railway lines that took place in the years following the Russo-Japanese War of 1904–1905, however, the government share of both cargo and passenger transportation rose to 80 %.

Chapter 3, by Ueda, is a methodological survey of the history of ecology and the environment in China. Since the terms used to express the concepts "ecology" and "environment" in different languages necessarily developed in different historical and cultural contexts, it is not surprising that their nuances in China and Japan differ from their nuances in Europe and the United States. In China, the terms used to describe "ecology" and the "environment" entered via Japan. In 1980, a term equivalent in meaning to "ecological economics" appeared. In the 1990s, reflecting the spread of pollution in China itself and the growing awareness of environmental problems all over the world, a term equivalent to "ecological environment" was also created. After examining this process, the author suggests the need for a new field of "socio-ecological history" that will bring together socio-economic and ecological history.

With Chap. 4, by Akatsu, we turn to Japanese research into European economic history. He analyses parliamentary debates in order to show how the British Smoke Nuisance Abatement Act of 1821 incorporated the interests of both politicians, landlords, and industrialists. It is not surprising that the act was promoted by landowners and the urban propertied classes, since they were damaged by air pollution. Factory owners in some industries responsible for pollution, such as textiles and food, also supported it. They hoped that the act would reduce energy costs by introducing new smoke prevention technologies. The act of 1821 did not itself establish any regulations against pollution. Even so, it signaled recognition by both government and parliament that air pollution was something that should be regulated by law. In that sense, it laid the basis for more interventionist legislation in the 1840s.

In Japan, high-quality research monographs and articles are published every year. Yet, while scholarly works originally published in European languages are often translated into Japanese, it is only in very rare cases that the reverse occurs. This has led to a regrettable knowledge gap for scholars who are unable to read Japanese. In an effort to address this imbalance, this volume also introduces reviews of recent academic works published in Japanese as well as a review by a Japanese scholar of an English language book that has recently been translated.

The four reviews are all evaluations of books on energy and the environment. They are a history of environmental problems in Tokugawa Japan by Andō Seiichi (Chap. 5); a study of the development of the Japanese electric power industry by

Kikkawa Takeo (Chap. 6); a volume on energy and corporate activities in modern Japan edited by Ogino Yoshihiro (Chap. 7); and a translation of J. R. McNeill's *Something New under the Sun* (Chap. 8).

Finally, I must thank Dr Koichi Inaba who translated all the book reviews, Dr Jeff Kurashige who translated Chap. 3, and Ms Ruth Fallon and Ms Louisa Rubinfien for checking and improving the drafts.

Shinjuku-ku, Tokyo, Japan S. Sugiyama

Contents

Part I Perspectives on the History of Energy and the Environment in Japan, China and Britain

1 **From Firewood to Coal: Deforestation and the Development of the Silk Reeling Industry in Modern Japan** 3
S. Sugiyama and Izumi Yamada

2 **The Government Railways and the Procurement of Railway Sleepers in Prewar Japan** 31
Asuka Yamaguchi

3 **The History of Ecological Environment: Ideas Derived from Chinese Research** 69
Makoto Ueda

4 **The Problem of Air Pollution During the Industrial Revolution: A Reconsideration of the Enactment of the Smoke Nuisance Abatement Act of 1821** 85
Masahiko Akatsu

Part II Book Reviews

5 **Review of Seiichi Andō, *Kinsei Kōgaishi no Kenkyū* (The History of Environmental Pollution in Early Modern Japan)** 113
Yuko Okada

6 **Review of Takeo Kikkawa, *Nihon Denryokugyō Hatten no Dainamizumu* (Dynamism in the Development of Japan's Electric Power Industry)** 117
Ryoshin Minami

7 Review of Yoshihiro Ogino (ed.) *Kindai Nihon no Enerugī*
   to Kigyō Katsudō: Hokubu Kyūshū Chiiki o Chūshin to Shite
   (Energy and Corporate Activities in Modern Japan: The
   Case of Northern Kyūshū) 123
   Naoki Tanaka

8 Review of J. R. McNeill, *Something New Under the Sun:*
   An Environmental History of the Twentieth-Century World
   (20 Seiki Kankyō Shi. Translated by Masatomo Umitsu
   and Tsunetoshi Mizoguchi) 127
   Shoko Mizuno

Index 133

Contributors

S. Sugiyama is a professor emeritus at Keio University and was president of the Socio-Economic History Society, Japan, between 2011 and 2014. He obtained a Ph.D. in history from the University of London in 1981. His main research field is the economic history of Japan and Asia. His publications include *Nihon Keizaishi* (An Economic History of Japan) (Iwanami Shoten, 2012) and *Japan's Industrialization in the World Economy 1859–1899* (Athlone, 1988).

Izumi Yamada graduated from the Faculty of Economics, Keio University, and is presently an employee of Sakata Seeds Corporation.

Asuka Yamaguchi is an assistant professor of economics at Nagoya City University. She obtained a Ph.D. in economics from Keio University in 2011. Her main research field is the economic history of Japan. Her publications include *Shinrin Shigen no Keizaishi: Kindai Nihon no Sangyōka to Mokuzai* (An Environmental Economic History of Forest Resources: Timber and industrialization in Modern Japan) (Keiō Daigaku Shuppankai, 2015).

Makoto Ueda is a professor of history at Rikkyo University. He obtained an M.A. in humanities from the University of Tokyo in 1982. His main research field is the social history of China. His publications include *Tora ga Kataru Chūgokushi* (Chinese History as narrated by a Tiger) (Yamakawa Shuppansha, 1999) and *Umi to Teikoku* (Seas and Empires: A History of the Ming and Qing Eras) (Kōdansha, 2005).

Masahiko Akatsu is a lecturer in economic history at Meiji University. He obtained a Ph.D. in economics from Meiji University in 2007. His main research field is the economic history of modern Britain. His publications include "Jūkyū Seiki Chūyō no Igirisu ni okeru Taikiosen Mondai" (The Air Pollution Problem in mid-19th Century Britain: The Smoke Prohibition Bill of 1844) in *Rekishi to Keizai* (The Journal of Political Economy and Economic History), 47–4.

Yuko Okada is an associate professor of economics at Kyushu Kyoritsu University. He obtained an M.A. in commerce from Waseda University in 1988. His main research field is the economic and business history of Japan. His publications include "Kosaka Kōzan Engai Mondai to Hantai Undō, 1901–1917" (The Movement against Smoke Pollution in the Kosaka Copper Mine), *Shakai Keizai Shigaku* (Socio-Economic History), 56–3.

Ryoshin Minami is a professor emeritus at Hitotsubashi University. He obtained a Ph.D. in economics from Hitotsubashi University. His main research field is the theory of economic development. His publications include *The Turning Point in Economic Development* (Kinokuniya, 1973) and *Nihon no Keizai Hatten* (Economic Development of Japan) (Tōyō Keizai Shinpōsha, 1981).

Naoki Tanaka is a professor emeritus at Nihon University. He obtained a Ph.D. in economics from Kyushu University. His main research field is the economic history of Japan. His publications include *Kindai Nihon Tankō Rōdōshi Kenkyū* (Studies in the Labour History of Coal Mines in Modern Japan) (Sōfūkan, 1984).

Shoko Mizuno is a professor of economics at Shimonoseki City University. She obtained a Ph.D. in humanities from Osaka University in 2002. Her main research field is the environmental history of the British Empire. Her publications include *Igirisu Teikoku kara miru Kankyōshi: Indo Shihai to Shinrin Hogo* (An Environmental History of the British Empire: Forest Conservation in Colonial India) (Iwanami Shoten, 2006).

Part I
Perspectives on the History of Energy and the Environment in Japan, China and Britain

Chapter 1
From Firewood to Coal: Deforestation and the Development of the Silk Reeling Industry in Modern Japan

S. Sugiyama and Izumi Yamada

Abstract The purpose of this article is to study the link between deforestation and the development of the silk reeling industry in Suwa district in Nagano prefecture from the 1870s to the 1900s, and the subsequent shift from firewood to coal. Since the Tokugawa period, firewood for the silk industry had come from land held in common by several villages. From the late 1870s, the development of silk as the most important export industry produced a shortage of firewood; by the mid-1880s, traditional sources were being supplemented by the transfer to silk producers of trees on government land. An attempt by the prefectural government to encourage tree-planting was unsuccessful, and in the early 1890s it became necessary to transport firewood from neighbouring districts.

The steam boilers which had been used in the silk industry since the late 1870s were cheap to buy, but too weak for use with coal. Coal, therefore, did not become important until around the turn of the twentieth century, when steam boilers had become stronger and the gap in the relative prices of coal and firewood had been reduced.

Keywords Firewood • Coal • Commons • Deforestation • Silk reeling

This chapter is a revised version of S. Sugiyama and Izumi Yamada, "From Firewood to Coal: Deforestation and the Development of the Silk Reeling Industry in Modern Japan" in Yano (2008). It originally appeared in *Shakai Keizai Shigaku* 65(2) (July 1999), pp. 3–23, as a Japanese-language article entitled "Seishigyō no Hatten to Nenryō Mondai: Kindai Suwa no Kankyō Keizaishi" (The Problem of Fuel: an Economic History of the Environment with reference to the Nagano Silk Reeling Industry)

S. Sugiyama (✉)
Keio University, 2-15-45 Mita, Minato-ku, Tokyo 108-8345, Japan
e-mail: sugiyama@z8.keio.jp

I. Yamada
Sakata Seed Corporation, Yokohama, Japan
e-mail: i-kawai@sakata-seed.co.jp

© Socio-Economic History Society, Japan 2015
S. Sugiyama (ed.), *Economic History of Energy and Environment*, Monograph Series of the Socio-Economic History Society, Japan, DOI 10.1007/978-4-431-55507-0_1

1 Introduction

It is beyond doubt that while industrial development has raised income levels and living standards, it has also brought about worldwide environmental destruction. As a result it is imperative to find a balance between economic development and the conservation of the environment. These social circumstances have drawn the attention of economic historians to environmental history. Their aim is to clarify the relationship of the environment to both industrial development and human economic activity in general.[1]

The purpose of this article is to examine the shift from firewood to coal in the Japanese silk reeling industry from the viewpoint of the economic history of the environment.[2] The focus will be on Suwa, a district in the mid-Japan prefecture of Nagano, during the rapid economic changes that occurred from the 1870s to the 1900s. Suwa was a centre of the silk reeling industry, a major export industry for Japan during the late nineteenth and early twentieth centuries Japan. This article makes extensive use of prefectural and local documents.

As an inter-disciplinary field, environmental history shares much common ground with other academic fields. In Japan, scholars have mainly concentrated on the history of pollution through examining court cases related to the heavy and chemical industries. Examples are the Ashio copper mine, the Kamioka silver and zinc mine, and the Chisso factory in Minamata (Oda 1983; Kamioka 1984). Recent publications by Andō Seiichi (1992) and Conrad Totman (1989) have widened the perspective by drawing attention to pre-modern Japan. Particularly important is the work by Chiba Tokuji in the area of economic geography. This has revealed the relationship between ecological change and the development of the salt and pottery industries (Chiba 1991).

"The tragedy of the commons" is a frequent topic in the fields of development economics and environmental economics.[3] However, while there are many case studies on the economic functions of Japanese village communities including the role of common land (Nakamura 1956; Nakamura et al. 1962), they have tended to ignore this issue. On the other hand, scholars of the history of Japanese land systems have overwhelmingly condemned the Meiji government for suppressing the traditional rights shared by neighbouring farmers when it disassembled common land and reorganized it into state land. By contrast, this article will suggest that deforestation did occur in common lands if there was a high demand for firewood

---

[1] For recent studies see, for instance, Crosby (1986), Ponting (1992), Worster (1993), Elvin and Liu (1998), and Pomeranz (2000).

[2] Wrigley (1988) argues that dependence on coal and oil for energy occurred by chance rather than out of necessity, but we do not agree with this view.

[3] Up until now, there has been a tendency to emphasize the good management of common land in pre-war Japan, including the Tokugawa period (Hayami 1995). However, the evidence presented in our study suggests that, if economic demand for resources was high, the mechanisms of common land management were unable to prevent environmental damage.

from industry, and that management systems did not work effectively in such cases. In addition, it should be noted that disassembling of the commons took place very gradually.

A number of studies have examined the silk reeling industry in Suwa (Yagi 1960; Kitajima 1970), but the main focus has been on the factors that made Japanese raw silk an internationally competitive export article. More specifically, research has covered the financing that made the purchase of cocoons possible, since this occupied 70–80 % of the production costs (Yamaguchi 1966; Ishii 1972); the management and organization of individual silk firms (Hirano 1990; Matsumura 1992); labour markets and the labour conditions of the female workers including the wage system (Ōishi 1968; Ishii 1972; Tōjō 1990); and the expansion of, and technological progress in, sericulture and silk reeling (Kiyokawa 1995; Suzuki 1996).

The issue of fuel has not received much attention. One reason for this is that fuel only accounted for around 3 % of the total production cost.[4] Nevertheless, the availability of fuel such as firewood and charcoal was a crucial factor in the development of the silk reeling industry. In fact, a classic study of the history of the silk industry in Nagano edited by Eguchi Zenji and Hidaka Yasoshichi divides the development of the silk reeling industry in the prefecture into three periods according to changes in the fuel source: a first period, to around 1892, when firewood was the main source of fuel; a second period, from around 1892 to around the time of the Russo-Japanese War (1904–1905), when both firewood and coal were used; and a third period, after the Russo-Japanese war, when coal took over as the main source (Eguchi and Hidaka 1937, vol. 3). The official history of Hirano village, the centre of the Suwa silk reeling industry, refers to the fuel problem, but merely lists the facts without further investigation (Hiranomura 1932, vol. 2).

2 Deforestation and the Development of the Silk Reeling Industry

2.1 Deforestation and Reforestation in Nagano Prefecture

First, it is necessary to provide an overview of energy sources in pre-war Japan. The average share of firewood and charcoal in the total energy supply was 90 % for 1880–84, 75 % for 1895–99, 50 % for 1905–09, and 37 % for 1910–14. Coal gradually replaced firewood and charcoal as the most important energy source, reaching 58 % in 1910–14, but firewood and charcoal persisted as the main energy source for pre-war Japanese households (Umemura et al. 1966; Makino 1996).

---

[4]Firewood and charcoal accounted for about 3 % of the total production expenditure, less than the expenditure on cocoons and workers' salaries (Naganoken 1980).

Since the Tokugawa period, the main source of firewood had been land (primarily forest-covered mountains) held in common by neighbouring villages. Deforestation in these common lands was already emerging as a problem in the early 1870s. At first the Meiji government encouraged the cultivation and privatization of government property (the land previously held by the feudal domains). But the spread of indiscriminate felling of trees led the government to change its attitude to privatization in 1873, and to define the division between state and private land in the following year. In 1876, the government surveyed the remaining customs regarding the use of common land; after 1878, many commons were integrated into state property (Okayashi 1976; Tsutsui 1978).

In 1884, the journal of the Forestry Bureau reported that deforestation and the general depletion of forest resources was a result of the deregulation of previously feudally owned territory, the development of mining and other industries, the rapid increase in the demand for construction materials, and other factors such as theft. At the same time, however, it pointed out that once land was integrated into state property residents experienced difficulties, particularly in obtaining firewood. The Bureau specifically linked deforestation in the three major silk reeling prefectures of Nagano, Gunma and Yamanashi to factors closely related to the development of the silk industry, including the increasing demand for firewood and construction materials (Tsutsui 1978). The patterns of deforestation found in Suwa were therefore broadly similar to those found in major silk reeling regions in other parts of Japan.

As a result of the development of the silk reeling industry in Nagano prefecture, deforestation of mountain land had proceeded to a considerable degree by the end of the nineteenth century. A contemporary author remarked on the vista of mountains bereft of trees stretching from the peak of Shiojiri pass to Suwa (Shinano Sanrinkai, no. 10, February 1914; Chiba 1953). Similar deforestation was observed in the northern part of Nagano prefecture. By the late 1880s, it was reported that "the abundant firewood forest of this region had been cut down and the surface of the mountainsides can now be seen" (Eguchi and Hidaka 1937, vol. 3, p. 971). Reporting in 1899 on the forests of the prefecture, Sonoyama Isamu, the governor, stated that deforestation had reached a peak, and emphasized the direct and indirect effects on national security and industry of this destruction of the balance of the environment.[5]

During the Meiji period, forests were largely divided into imperial, state, public, and individual private ownership. Public forest land was owned by entities such as prefectures, counties, municipalities, and villages. In 1901 it was reported that half of the 546,083 ha of state-owned forest land in Nagano prefecture was of no use due to difficult access, while another quarter was in ruins. The forest land available for use therefore amounted to only one quarter, 136,500 ha. Similarly, the combined

---

[5]Shinano Sanrinkai, no. 1, November 1902. From the late 1870s, water mills were built along the Tenryū river in order to provide power for silk reeling. However, they impeded the flow of water and caused Lake Suwa to flood. This started a dispute between the villages around Lake Suwa and the villages along the Tenryū river (Horie 1930; Takamura 1995).

total of public and private forest land was 352,066 ha, but "a quarter was located in an area so difficult for transportation and delivery that it was of no use, and a half of the remainder was so devastated that almost no large trees could grow." The report continued that "the amount of forest land that can be used at this point is about a third of the total private area", in other words 117,400 ha. Of the private forests, "the majority of those owned by individuals are located in areas suited for transportation and delivery; hence their forestry products are in high demand and profitable, which promotes both tree-felling and tree-planting." On the other hand, public forests had "deep-rooted traditions and customs which are hard to alter, so that … attempts to start reforestation are extremely sporadic. Tree-felling continues and mountains are despoiled, but no effective measures are taken." Ultimately, "despite the spread of unrestricted tree-felling, no attempts are made to change old customs or practices involving old methods, leaving the forests to be cut down as freely as if they were the personal property of every villager." Thus village forests (common land) were the most vulnerable (Shinano Sanrinkai, no. 2, December 1903).

Faced with such nationwide deforestation, in 1878 the Meiji government attempted to encourage tree-planting in state land, including mountains. The prefectural government of Nagano introduced subsidies for tree-planting, but to no effect (Ōi 1973a, b). Substantial measures for reforestation were not introduced until 1897, when *Shinrinhō* (The Forest Law) was issued to regulate tree-felling and planting in badly affected common land. In 1899, the Nagano prefectural government also issued regulations to control the management of forest land owned by municipalities and villages, and to grant subsidies to tree planters to encourage the reforestation of public forests. In 1904, the prefectural government attempted to prevent further deforestation and promote reforestation by altering the customs of common land and incorporating village forests into the basic assets of municipalities (Shinano Sanrinkai, nos 1, 10 and 19; Matsunami 1919; Okayashi 1976). It understood the need to prohibit tree-felling for environmental reasons; however, since forbidding a custom that was considered part of the traditional rights of farmers would have a drastic effect on people's ability to obtain essential supplies of firewood, it was difficult for the authorities to introduce effective changes (Naganoken 1986).

2.2   Forest Land in Suwa

As was stated above, since the Tokugawa period firewood had come from land that was held in common by neighbouring villages. Farmers had the right to enter commons such as Mt Yokokawa, Mt Higashi and Mt Tozawa to collect wood and grass. For instance, the 2,671 ha comprising the public land of Mt Yokokawa was jointly owned by the five adjacent villages, including Hirano. Each village had different rights concerning entrance routes, purposes of use, and seasons when use was permitted. Tree-felling was initially prohibited; however, it gradually increased as the disassembling of forest policy under the declining feudal domain system

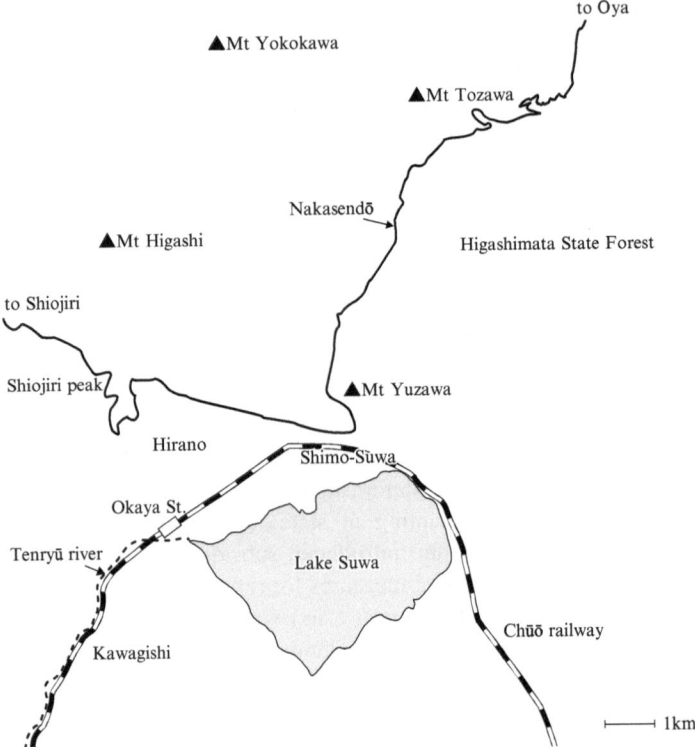

Fig. 1.1 Map of Suwa

led to a loosening of restrictions. A part of Mt Higashi which had been common land for villages in both Chikuma and Suwa since the mid-seventeenth century was registered as public land in 1872, but was turned into state land in 1878. There were many disputes between the two regions over the borders of the common land. This continued until the official division of Mt Higashi in 1909 (Hiranomura 1932, vol. 1; Fukushima et al. 1956, 1958; Okayashi 1976).

Table 1.1 shows the situation of forest land in Suwa at the end of 1903. At this time there were no state forests in Suwa county except a small imperial forest. Public forest accounted for 58.2 % of all forest land, 41.6 % was private forest, and the small remaining amount was held by shrines and temples. This pattern was characteristic of Nagano prefecture as a whole, but a significant characteristic of Suwa was the high percentage of land covered by bushes rather than trees in both public and private forests. While the overall average percentage of land not covered by trees was 36.7 % for public forest and 43.0 % for private forest, in Suwa county it was 61.2 % and 56.6 % respectively. This indicates that deforestation was more advanced in Suwa, particularly in public forests. Furthermore, in Suwa county conifers such as red pine and larch were predominant, while Japanese cedars, commonly seen in other areas of Nagano prefecture, were relatively rare (Shinano

Table 1.1 Public and private land in Suwa at the end of 1903 by ownership and nature of forestation (in hectares)

	Type of tree	Public	Shrines and temples	Private	Total
Land with trees					
Conifer	Red Pine	593	48	2,232	2,873
	Larch	957	5	1,107	2,069
	Others	1,626	7	516	2,121
Deciduous	Chestnut	679	1	319	999
	Japanese Oak	1,192	0	265	1,458
	Others	1,656	5	924	2,585
Sub-total		6,703	66	5,361	12,130
Land without trees		10,558	1	6,983	17,542
Total		17,261	67	12,343	29,672
		58.2 %	0.2 %	41.6 %	100 %
Percentage of land without trees		61.2 %	1.3 %	56.6 %	59.1 %

Source: Shinano Sanrinkai (1904), pp. 12, 21
Note: Totals are not necessarily consistent because of rounded figures

Sanrinkai 1904; Shinano Sanrinkai, no. 2).[6] This indicates that the purpose of reforestation in Suwa was to secure fuel sources, since red pine and larch had a relatively high energy efficiency.

Table 1.2 shows the area of public and private forest land in Suwa with details of both vegetation and ownership. In the mid-1880s, over three-quarters of all forest land consisted of mountains covered only by grass or bushes. This suggests that extensive and indiscriminate tree-felling had already taken place. Deforested areas decreased in parallel with the advance of reforestation, but more than 60 % of forest land was without trees throughout the late nineteenth century. 70–80 % of the combined total public and private forests were mostly commons jointly owned by villages; 70–80 % of this land was only covered by grass or shrubs. This indicates that village-owned forest land was the most affected by deforestation. According to Hirano's official records, in 1904, only 55 ha (2.0 %) of the total area of 2,726 ha owned by the village maintained a forest-like appearance (HYb). In 1911 a forestry bulletin remarked about village-owned forests that people were "gathering grass and firewood with no thought of tomorrow so that they have laid waste the mountain areas that are under public ownership" (Shinano Sanrinkai, no. 8, May 1911, p. 3). A 1913 survey of deforestation noted that the affected areas were not adequately

---

[6]The use of conifers in reforestation was highly criticized by the officials involved in forest and agricultural policy. They emphasized the need to restore soil fertility by planting deciduous trees first (Shinano Sanrinkai, nos 9 and 10).

Table 1.2 Public and private forest land in Suwa by ownership and nature of vegetation, 1884–1899 (in hectares)

Year	Total area	Vegetation		Percentage without trees (%)	Ownership		Percentage of joint ownership (%)	Percentage of common land without trees (%)
		Forest	Grass		Single	Joint (commons)		
1884	23,198	6,112	17,086	73.7	6,784	16,413	70.8	81.8
1885	23,487	6,186	17,301	73.7	6,872	16,615	70.7	85.3
1886	23,399	5,906	17,494	74.8	5,119	18,280	78.1	86.5
1887	22,889	8,056	14,834	64.8	6,375	16,516	72.2	80.4
1888	26,180	8,903	17,276	66.0	9,532	16,648	63.6	79.1
1889	29,581	11,256	18,326	61.9	5,751	23,831	80.6	68.3
1890	27,580	10,077	17,504	63.5	5,034	22,546	81.7	69.6
1891	30,701	10,298	20,404	66.5	6,023	24,678	80.4	72.8
1892	25,349	10,299	15,051	59.4	5,561	19,787	78.1	66.3
1893	30,683	10,297	20,385	66.4	6,434	24,249	79.0	76.1
1894	28,137	10,301	17,837	63.4	6,460	21,677	77.0	73.2
1895	28,156	10,309	17,848	63.4	5,348	22,809	81.0	69.5
1896	27,775	10,298	17,477	62.9	8,263	19,512	70.3	79.8
1897	27,127	10,306	16,821	62.0	8,262	18,866	69.5	79.1
1898	26,208	10,584	15,623	59.6	10,390	15,818	60.4	81.6
1899	24,062	11,390	12,671	52.7	8,232	15,830	65.8	70.3

Sources: Naganoken (NT). Figures for 1889 were revised by Naganoken Kangyōka (NKN)

Note: Figures for forest land include mountainous areas

Table 1.3 Details of tree planting in Suwa, 1882–1910

Year	Area (in hectares)	Number of trees
1882–1885	52	259,384
1886–1890	57	216,186
1891–1895	462	507,252
1896–1900	520	616,853
1901–1905	497	1,505,968
1906–1910	829	3,283,993

Sources: 1882, Naganoken Kangyōka (NKG); 1883–1896, Naganoken Kangyōka (NKN); 1897–1911, Naganoken (NT)
Note: Figures for the years 1886, 1890 and 1902 are not available

supervised because of the complicated ownership patterns, "the absence of any managing body, and the villagers' lack of community spirit" (Okayashi 1976, p. 255).[7]

In Suwa, reforestation was encouraged through local government subsidies from 1881. Table 1.3 shows the area of reforestation and the total numbers of trees planted after 1882. In Hirano, a total of 68,482 trees (mainly larch) were planted in an area of 34.5 ha during the period from 1881 to 1900. In the case of Mt Yokokawa, a total of 160,100 trees including larch were planted during the period from 1898 to 1904 (Table 1.4). However, the area of reforestation was estimated to cover only 3 % of the total common land (2,671 ha) of Mt Yokokawa.

3 The Supply and Demand Relationship of Firewood and Charcoal in Suwa

3.1 The Power Requirements of Cocoon Steaming and Silk Reeling

From the mid-1870s, the increasing U.S. demand for silk, driven by the rapid growth of its silk textile industry, encouraged the development of the silk industry in Suwa. Raw silk production increased after the introduction of improved traditional wooden reeling machines using water power to reduce production costs. Raw silk became Japan's major export article. Exports to the United States increased rapidly after 1882, replacing Chinese silk in the U.S. market, and overtaking Japanese exports to France in 1884 (Sugiyama 1988).

Table 1.5 shows the changes in the types of power used for cocoon steaming and silk reeling in Suwa in general and Hirano in particular. From the period of traditional sedentary reeling to the early stages in the introduction of filatures,

---

[7]As late as 1917, 92 % of the total forest of 25,385 ha in Suwa was still owned by villages (Naganoken Suwagun Yakusho 1918).

Table 1.4 Tree planting in Hirano and Mt Yokokawa, 1881–1904

Year	Hirano Village			Mt Yokokawa		
	Types of tree	Area (in hectares)	No. of trees	Types of tree	No. of trees	Price (in yen)
1881	Larch	1.25	4,026			
1882	Larch	1.03	3,360			
1883	Larch	0.81	2,794			
1884	Larch	1.06	3,200			
1885	Larch	0.53	7,964	Larch	n.a.	
1889	Larch	2.19	10,000			
1890	Larch	2.19	10,000			
1891	Larch	2.19	10,000			
1892	–	–	–			
1893	–	–	–			
1894	Larch	0.27	344			
1895	Larch	0.27	344			
1896	Larch	9.22	5,450			
1897	–	–	–			
1898	–	–	–	Larch	5,000	7.50
1899	Pine	9.92	5,000	Larch	3,000	6.90
1900	Larch	11.90	6,000	Larch	10,000	16.50
1901				Larch	108,100	66.25
1902				Larch	3,000	3.60
1903				Larch and Cedar	21,000	43.80
1904				Larch	10,000	15.00
Total		34.81	68,482		160,100	159.55

Sources: For Hirano Village, Hiranomura and Kawagishimura Kangyō Kakari (1885); Hiranomura Yakuba (HYa), (HYb). For Mt Yokokawa, Hiranomura Yakuba (1906); Hiranomura Yakuba (HYf)

firewood and charcoal were used to steam cocoons and keep the water hot. Firewood was the main source of fuel until coal was introduced in the mid-1890s; however, there were no major differences in the use of firewood between steam boilers and the other two boiling methods because steam boilers were originally heated by firewood (Hiranomura 1932, vol. 2; Eguchi and Hidaka 1937, vol. 3). There were two types of steam boiler, one for heating and the other to provide power for turning reeling machines. Steam boilers for heating were gradually replaced by boilers that could both heat and provide power. While the use of heating boilers for cocoon steaming grew rapidly from the late 1870s to the mid-1880s, water power was widely used to turn reeling machines because the availability of water power from the Tenryū river gave many silk reeling areas in Suwa a natural advantage (Takamura 1995). By the 1900s, steam boilers were playing a major role in Hirano because it did not have good access to the river, but water power remained dominant in Suwa as a whole. It was not until the second half of the 1900s that steam boilers became a widespread source of power.

Table 1.5 The silk reeling industry in Suwa and Hirano, 1878–1905

Year		1878	1883	1885	1886	1887	1893	1896	1900	1905
Suwa										
No. of factories		108		139	113	110	219	39	35	115
No. of basins				3,571	3,245	3,650	9,338	11,607	10,109	13,731
Method of boiling cocoons	Steam	51		122	92	109	219	39	35	114
	Charcoal	57		17	21	1	0	0	0	1
Power for reeling	Steam			0	0	0	24	7	2	38
	Water			117	90	86	146	31	26	60
	Manual			22	23	24	49	1	0	7
	Steam and water			0	0	0	0	0	6	9
	Others			0	0	0	0	0	1	1
Hirano										
No. of factories		28	43	53	50	65	91	16	14	46
No. of basins		940	1,234	1,399	1,386	1,755	4,543	6,741	5,877	7,583
Method of boiling cocoons	Steam	0	17				91	16	14	46
	Charcoal	28	26				0	0	0	0
Power for reeling	Steam		0				23	4	1	34
	Water		31				34	11	6	5
	Manual		12				34	1	1	0
	Steam and water		0				0	0	4	7
	Others		0				0	0	2	0

Sources: 1878 and 1883, Naganoken Kangyōka (NKN) 1885; Eguchi and Hidaka (1937), vol. 2, p. 646; 1886 and 1887, Naganoken Sanshigyō Kumiai (1887, 1888, 1893–1905), Nōshōmushō Nōmukyoku, 1 (1895), 2 (1898), 3 (1902), and 4 (1907). Figures for the number of factories in Hirano village between 1885 and 1887, and for the number of reeling basins in Hirano village between 1878 and 1887, are based on Hiranomura (1932), vol. 2, pp. 273–274
Notes: For the period 1893–1905, figures for filatures are for those with more than ten reelers. The number of factories for 1896 and 1900 refer to silk associations only. Figures for the number of factories in 1905 give the total of filatures and silk associations excluding re-reeling factories

The quality of the materials used to make steam boilers differed according to their purpose. Power boilers used higher pressured steam than heating boilers, so they needed to be stronger. Cast-iron heating boilers were introduced into Suwa from the 1870s because they were cheap and economical. They spread rapidly among the smaller silk reeling factories. For technological reasons it was difficult to produce them in large sizes. However, in 1877 Maruyama Ironworks successfully produced a heating boiler made of thin iron plate that was only 0.3 cm thick. With a higher level of efficiency than the cast-iron boiler for a similar price, this new type of boiler spread rapidly among small and medium-sized silk factories. Maruyama Ironworks also began to produce a boiler with several pipes that had improved heat efficiency. Against the background of the increasing overseas demand for raw silk, many nearby ironworks began to supply this new type of boiler. It became the standard model and facilitated an expansion in the size of silk reeling factories. To prevent explosions, the boilers were given water gauges, then manometers and finally safety valves, but problems related to structure and the quality of materials were not completely overcome. Nevertheless, and even paradoxically, these boilers spread widely. This was because their thin iron plate structure responded well to the low heating power of firewood, and because the calking to prevent leakage from the joints was so weak that explosions from high heat pressure were unlikely. But once coal, with its high heating efficiency, emerged as a power source for reeling, the weaknesses of thin iron plate boilers made them redundant (Hiranomura 1932, vol. 2; Suzuki 1996).

3.2   The Supply and Demand Relationship of Firewood and Charcoal Around 1880

Figures for the supply and demand of firewood and charcoal include not only the amount consumed by the silk reeling industry but also that for other industries and for household use. In addition, the supplies came from common land. These factors make it difficult to give precise figures for supply and demand.[8] Table 1.6 shows the consumption of firewood and charcoal for Suwa in general and Hirano in particular, but since firewood was not used solely for silk reeling, the figures for consumption per basin in Table 1.6 do not necessarily indicate exact quantities. The figures for firewood consumption per basin in Hirano seem high, but this is because of Hirano's high level of involvement in silk reeling and the fact that it could not use water power. Firewood and charcoal consumption increased dramatically from the late 1880s to the early 1890s, a period which also saw a rapid increase in raw silk production. Firewood consumption per basin decreased in the mid-1880s, but

---

[8]Industries that consumed firewood in Nagano other than silk reeling were brewing, lime manufacturing and hemp manufacturing. In 1901 the silk reeling industry accounted for 79 % of the total consumption of firewood and charcoal (Shinano Sanrinkai, no. 2).

Table 1.6 Consumption of firewood and charcoal for silk reeling in Suwa and Hirano, 1879–1903

	Suwa County				Hirano Village						
Year	Raw silk production (in tons)	No. of basins	Consumption of firewood (in tons)	Consumption of firewood per basin (in tons)	Raw silk production (in tons)	No. of basins	Consumption of firewood (in tons)	Price of firewood per ton (in yen)	Consumption of firewood per basin (in tons)	Consumption of charcoal (in tons)	Price of charcoal per ton (in yen)
1879	24,979	1,369	2,689*	1.97*	6,143	655	1,286	3.68	1.97	13	11.44
1885	67,376	2,242	3,634*	1.62*	32,040	1,399	2,269	6.21	1.62	–	–
1888	156,131	4,234	13,909*	3.29*	111,945	2,192	7,200	2.51	3.29	44	10.19
1890	262,826	8,097	28,845*	3.56	143,336	3,362	11,970*	2.53	3.56*	–	–
1892	384,923	8,420	40,500	4.81*	215,966	3,977	17,018*	2.96	2.71	–	–
1893	412,999	10,883	55,298*	5.08*	227,160	4,764	24,206	2.93	5.08	208	10.67
1898	385,639	9,969	57,795*	5.80*	219,934	4,821	27,953	4.00	5.80	300	18.91
1903	654,323	14,495	66,533*	4.59*	404,063	6,537	30,000	10.67	4.59	300	21.33

Sources: Eguchi and Hidaka (1937), vol. 3, pp. 968–970, 1138–1139; Hiranomura (1932), vol. 2, pp. 273–274, 384, 386; Hiranomura Yakuba (HYe)

Notes: * estimates. Figures for raw silk production and no. of basins for Suwa in 1879 are those of 1881. Figures for raw silk production of Hirano in 1888 are those for 1889. Figures for the price per ton of Hirano for 1885 and 1890 are those at Matsumoto (See Fig. 1.2)

this is probably because of the increased efficiency of steam boilers. The increase in firewood consumption per basin in the early 1890s was caused by the expansion in the size of factories through the formation of silk reeling associations, by the increase in consumption for other purposes, and by the spread of steam boilers requiring more firewood for both heating and power.

Table 1.7 shows the supply and demand relationship for firewood and charcoal in Hirano from 1879 to 1880. The consumption of all types of firewood, including pine, for silk reeling in Hirano village during this period is shown as 1,703 t, which is equivalent to 1,275 t per year (HYd 1875; Hiranomura 1932, vol. 2). It should be noted that in calculating these figures, an error was detected in *Hirano Sonshi*, a widely respected source. This claims that from June 1879 to July 1880 the demand for both pine and other firewood for silk reeling amounted to 3,194 t, totaling 10,028 yen. If these figures were correct, it would mean an anomalous situation of one piece of firewood weighing as much as 188 kg and costing as much as 0.5 yen. Furthermore, Nakayamasha, the only firm that used pine firewood for silk reeling, would have consumed 1,875 t during this period. This would mean that its firewood consumption per basin for the 14 months from June 1879 to July 1880 was the incredibly high figure of 19 t. On going back to the original record, it became clear that the editors of *Hirano Sonshi* had detected an error, but corrected it in the wrong way, by adding one zero to the value rather than removing one zero from the weight. This is corroborated by the fact that in the original record, the price for a single stick of pine firewood for sake brewing was given as 0.5 yen. It was therefore clear that the firewood consumption of Nakayamasha should be adjusted downwards to 187.5 t as in Table 1.7.

The supply of firewood and charcoal from Mt Yokokawa from July 1879 to June 1880 amounted to 188 t of pine firewood, 4,303 t of other miscellaneous firewood, and 169 t of mixed charcoal (Table 1.7). It should be noted that figures for Mt Yokokawa do not indicate the demand for firewood from Hirano alone, since it was held in common by Hirano and its neighbouring villages. The figures suggest that Hirano had to go elsewhere to satisfy its demand for pine firewood and oak charcoal, but that this was not necessary in the case of miscellaneous firewood.

Shimo-Suwa, another district in Suwa, appears to have been plagued by fuel shortages from early on. In 1877, Shimo-Suwa village applied for the transfer of trees from the state forest of Mt Yuzawa that had been felled by high winds. In 1878, representatives of the village jointly petitioned the prefectural government for the transfer of grass and bushes from the forest at reasonable prices, since no dead, fallen, or burned trees remained in the much nearer Higashimata state forest. In the following year, a silk producer of the village submitted a joint petition asking for the transfer of state forest land and trees at a reasonable price so that the size of the filatures could be increased. He also submitted a similar joint petition for the transfer of trees from the state forest of Mt Yuzawa, claiming that "the firewood will soon run out" and that they were "at a loss what to do since there is no prospect of obtaining any more" (Shimo-Suwa Chōshi Hensan Iinkai 1969, pp. 1226–1228, 1253–1254). The transfer of trees from state forests to local residents gradually became a common occurrence in the region.

Table 1.7 Supply and demand of firewood and charcoal at Hirano, 1879–1880

Fuel	Demand for firewood and charcoal at Hirano Village, June 1879 – July 1880				Firewood and charcoal obtained from Mt Yokokawa, July 1879 – June 1880			
	Quantity	Weight (in tons)	Value (in yen)	Price per ton (in yen)	Quantity (in piece)	Weight (in tons)	Price on the spot (in yen)	Price per ton (in yen)
By type of fuel								
Pine firewood	13,580 pieces	254.6	679.00	2.67	1,000	187.5	250.00	1.33
Other firewood	96,200 pieces	1,435.5	5,502.64	3.83	229,500	4,303.1	9,470.02	2.20
Oak charcoal	985 bales	12.9	147.75	11.43	–	–	–	–
Other charcoal					–	168.8	900.00	5.33
By use								
Pine firewood for silk reeling	10,000 pieces	187.5	500.00	2.67				
Other firewood for silk reeling	87,900 pieces	1,311.0	5,027.88	3.84				
Oak charcoal for silk reeling	985 bales	12.9	147.75	11.43				
Pine firewood for sake brewing	3,580 pieces	67.1	179.00	2.67				
Other charcoal for sake brewing	8.300 pieces	124.5	474.76	3.81				
Total		1,703.1	6,329.39	3.72	230,500	4,659.4	10,620.02	2.28

Source: Calculated from Hiranomura (HYe)

3.3   The Supply and Demand Relationship of Firewood and Charcoal in the Late 1880s

From the late 1870s, increasing demand from the silk reeling industry began to cause a shortage of firewood. The annual consumption of firewood by the Hirano silk reeling industry was estimated to be 2,269 t in 1885. It was reported that "year by year producers are facing more difficulties in obtaining firewood. ... They are afraid that firewood and charcoal shortages will endanger their business" (Hiranomura 1932, vol. 2, p. 384). By the mid-1880s, there was a clear trend for traditional sources of firewood to be supplemented by the transfer to silk producers of trees from state land.

Table 1.8 gives detailed figures for the 15 silk associations (with a total of 110 factories) located in Suwa from the 1887 version of the *Annual Report of the Agency for Regulating Silk Associations in Nagano Prefecture*. Some obvious errors can be found, but it is still an important source, since there are no other records for the consumption of firewood and charcoal per silk association during this period (Naganoken Sanshigyō Kumiai, no. 2 1888). Kaimeisha in Hirano was the largest consumer in 1887, accounting for 28 % of the total firewood consumption of Suwa.

It is nearly impossible to calculate figures for forest lumbering since there is little information about tree-felling on common lands. In fact, the only way to estimate levels of deforestation is to use the grass/land ratios given in Table 1.2. These suggest that the real crisis occurred after the mid-1880s. We also have evidence of lumbering operations in government forests from the example of Higashimata state forest near the village of Shimo-Suwa. In the mid-1880s, it was not unusual to see large trees being cut into logs of 91 cm in length which were then fitted with shafts for transportation to factories for use as firewood (Hiranomura 1932, vol. 2; Shimo-Suwa Chōshi Hensan Iinkai ed. 1963). During the period from 1884 to 1888, 188,236 trees, 95 % of the total number of trees felled at state forests in Suwa, were destined for use as firewood. In addition, during this period, theft was the main cause (at 49 %) for the loss or damage of 45,993 trees, above other causes such as fire or wind and snow. In 1888, when the firewood shortage was particularly serious, theft was almost the sole cause behind the 19,758 damaged trees (Naganoken, for 1884–1888).

3.4   The Supply and Demand Relationship of Firewood and Charcoal in the Late 1890s

By the end of the 1880s, the firewood shortage had reached a crisis point. In the 1890s, movements to petition for the transfer of state and imperial forests grew more frequent. At the same time, the transport of firewood from neighbouring districts

Table 1.8 Consumption of firewood and charcoal at filatures in Suwa, 1887

Silk associations	Year of establishment	No. of factories	Method of boiling	Power for reeling	No. of basins	Firewood consumption (in tons)	Price of firewood per ton (in yen)	Charcoal consumption (in tons)	Price of charcoal per ton (in yen)
Kaimeisha	1878	22	Steam	Water and manual	800	8,568	0.73	9	8.4
Shichiyōseisha	1877	10	Steam	Water	432	2,835	1.14	30	8.03
Gakosha	1877	10	Steam	Water	338	2,793	0.76	11	11.36
Kairyōsha	1873	13	Steam	Water and manual	434	2,681	0.99	15	9.15
Hiranosha	1875	9	Steam	Water	300	2,186	1.07	13	10.67
Kanayamasha	1882	7	Steam and charcoal	Water	149	2,153	0.66	5	8.03
Yajimasha	1874	11	Steam	Water	278	2,066	0.58	9	11.84
Tōeisha	1876	8	Steam	Water	185	1,907	0.57	11	6.77
Hakutsurusha	1876	5	Steam	Water	290	1,620	1.07	17	7.12
Kōshinsha	1876	2	Steam	Water	135	1,221	0.33	5	11.52
Kaiseisha	1881	4	Steam and charcoal	Water and manual	57	661	0.71	3	7.15
Ōbeisha	1877	2	Steam	Water	68	557	0.69	2	9.79
Seigyōsha	1875	3	Steam	Water and manual	64	434	0.77	1	8.88
Hōshinsha	1878	2	Steam	Water	58	244	1.14	3	8.85
Ryōzensha	1880	2	Steam	Manual	62	218	1.33	1	9.71
Total		110			3,650	30,143	0.80	133	8.91

Source: Naganoken Sanshigyō Kumiai (1888), pp. 267–268, 275–277
Note: Obvious misprints of figures in the original source have been corrected

increased, and the prefectural government made a second attempt to encourage tree-planting. Eventually, the shortage of firewood encouraged a shift to a new fuel source: coal.

There is more data on fuel consumption for the 1890s than for earlier periods (Table 1.6). A fuel survey of the Nagano silk reeling industry for 1890 estimated that annual firewood consumption amounted to 58,500 t, with annual firewood consumption per basin at 2.5 t. These figures are similar to those given by a survey conducted around 1888–1891 for 14 filatures with a total of 4,000 basins. However, while this survey assumes a total of 150 annual working days, in 1889 the eight filatures located in Hirano had 211 working days. Therefore, it is possible that the annual consumption was actually greater (Hiranomura 1932, vol. 2; Eguchi and Hidaka 1937, vol. 3). Another document, dated November 1892, states that the annual firewood consumption for the Suwa silk reeling industry was a third of its total consumption of firewood, that is 40,500 t (HYe). In this case the annual consumption per basin would be 4.8 t.

Table 1.9 shows the consumption of firewood and coal by the various silk reeling associations in Hirano as of October 1892. According to these figures, the annual firewood consumption per basin was 2.7 t, which is much less than the figures for Suwa as a whole. By this time, therefore, a switch to coal was beginning in the larger silk reeling associations such as Kaimeisha, Ryūjōkan and Kairyōsha that had been major consumers of firewood. In 1893, the total consumption of firewood by filatures in Hirano was 24,206 t and the annual firewood consumption in Suwa as a whole was estimated at 55,300 t (Table 1.6). These figures are consistent with those given in other documents (HYe). Figures for lumbering operations are not available for the early 1890s, but for the period from 1896 to 1900 the annual average of tree-felling for firewood in forest land in the environs of Hirano was 1,312 t, and in 1900 83,625 t of timber still remained (HYb).

In November 1892, Kaimeisha submitted a petition to the Imperial Forest Bureau for the annual transfer of trees amounting to 8,100 t from the imperial forest land in Kami-Ina for 10 years beginning in 1892. When this was not granted, Kaimeisha submitted the same petition in the following year. Up until then, Kaimeisha had somehow managed to obtain enough firewood, but "the consumption of firewood has risen to such levels that supplies have been exhausted" (HYe). Kaimeisha admitted that even transfers of 8,100 t per year would not be enough to guarantee security of fuel for its operations (HYe), but in any case this second petition was accepted. At the end of 1893 a joint venture of seven local silk associations, including Kaimeisha, Shin'eisha, Ryūjōkan, Hiranosha and Kairyōsha, began joint lumbering operations; however, a rapid rise in the price of firewood caused by natural disasters, combined with transportation difficulties and the opposition of local farmers, hindered the enterprise. Eventually, five of the associations withdrew, leaving only Kaimeisha and Shineisha (Hiranomura 1932, vol. 2).

In September 1895, Kaimeisha and Shin'eisha applied to establish Suwa Shintan Kabushiki Kaisha (the Suwa Firewood and Charcoal Co.) with an initial capital of 20,000 yen for the purpose of "procuring and marketing firewood needed for the

Table 1.9 Consumption of firewood and coal at filatures in Hirano as of October 1892

Firms	Year of establishment	Capital (in yen)	No. of basins	Raw silk production (in bales)	Firewood consumption (in tons)	Value of firewood (in yen)	Coal consumption (in tons)
Kaimeisha	1878	500,000	1,736	2,017	5,795	17,170	26.25
Ryūjōkan	1890	50,000	1,251	783	2,068	6,127	46.24
Kairyōsha	1882	55,000	–	658	1,752	5,190	24.90
Shin'eisha	1890	32,960	955	392	1,069	3,168	–
Hiranosha	1886	100,000	495	393	1,064	3,154	–
Meishinsha	1890	42,000	99	312	842	2,496	–
Nishi–Hakutsurusha	1891	60,000	–	235	635	1,880	–
Yajimasha	1884	14,000	253	227	613	1,816	–
Kanayamasha	1882	12,500	434	127	343	1,016	–
Total		866,460	5,223	5,144	14,181	42,017	97.39

Sources: Compiled from Hiranomura Yakuba (HYe); Nōshōmushō Nōmukyoku (1908); Naganoken Tōkeisho (NT) (1893)

production of raw silk". Capital was provided by leading silk entrepreneurs such as Ozawa Fukutarō and Katakura Kanetarō. The company operated for 10 years from 1896 to 1906 in cooperation with local villagers. As a consequence, the fuel shortage was mitigated to some extent, but the business did not continue beyond the 10 year initial contract (HYe; Hiranomura 1932, vol. 2). Meanwhile, a similar petition to the Imperial Forest Bureau made by four Hirano silk associations in 1894 had given the following bleak description of the local landscape: "every mountain bares its bones in unspeakable desolation; our hopes of obtaining firewood in the future are gone" (HYe).

As a result of this fuel shortage, firewood and other fuel had to be transported from neighbouring districts. Suwa records state that in 1892, 17,531 t of firewood valued at 51,425 yen were transported from the neighbouring districts of Kami-Ina and Higashi-Chikuma; 180 t of charcoal valued at 1,920 yen were delivered from Kami-Ina, Higashi-Chikuma and Chiisagata; and 854 t of coal valued at 5,237 yen from Higashi-Chikuma (HYa). Firewood amounted to 88 % of the total fuel deliveries. It was transported by wagon to Suwa, but the high transportation costs pushed up the price that was charged there (Naganoken ed. 1980). As a result, the following conclusions can be drawn regarding the supply and demand relationship of firewood in Hirano in the early 1890s. The total annual demand for firewood for silk reeling in Hirano amounted to 24,206 t. Of this, 1,312 t came from local tree-felling (taking an average of the figures for 1896–1900), 8,100 t from the transfer of trees from state forests in Suwa, and 17,531 t from deliveries from outside the region. In other words, Hirano was able to satisfy a third of its firewood needs with supplies from nearby common land and state forests, but had to rely on outside areas for the remaining two thirds.

The increasing shortage of firewood for fuel led to transports not only from districts within Nagano prefecture, but also from other prefectures. In 1897, 290 t of firewood valued at 1,630 yen were transported to Nagano from Gunma and Niigata prefectures, 1,307 t of charcoal valued at 22,715 yen from Gunma, Niigata and Yamanashi, and 3,805 t of coal valued at 37,502 yen from Tokyo and Hokkaidō (Naganoken 1897).

In Shimo-Suwa, so many trees in the commons and neighbouring mountains were cut down as firewood for boilers during the agricultural off-season that the mountains in the area had been almost completely stripped of trees by the early 1890s. A Shimo-Suwa silk association applied for the transfer of trees from Higashimata imperial forest around 1897, but local residents opposed this on the grounds that the imperial forest was indispensable for their supplies of water and for irrigation. As a result, the Imperial Forest Bureau rejected the application. In 1898 Mt Yuzawa was privatized, and eventually the land was divided into pieces for sale. Within a few years the pine forest had vanished. Even the tree stumps had been dug up for use as fuel. This forced the local assembly to ban the removal of tree stumps in 1900 (Shimo-Suwa Chōshi Hensan Iinkai ed. 1963).

4 The Shift from Firewood to Coal

In Suwa, coal use began around 1877 (Hiranomura 1932, vol. 2). At this stage consumption was limited, since coal was relatively expensive and proved dangerous to use. In addition to technological problems such as the fact that the boilers of that time were not suited to coal, as was mentioned before, there was a lack of skilled boilermen. Table 1.10 shows coal consumption in the private sector in Hirano in particular and Nagano as a whole after 1890. It is clear that coal consumption in Hirano and Nagano increased at parallel rates. In 1896, and after 1903, the increase was rapid, but there was stagnation in the late 1890s and a decline from 1900 to 1902. It can be assumed that these patterns were closely related to the availability of firewood as an alternative fuel, and to price fluctuations.

Matsumoto is the closest city to Suwa for which contemporary price details are available. Figure 1.2 shows the prices of firewood, charcoal and coal in Matsumoto. For purposes of comparison, prices are shown according to ratios of energy efficiency, with 1 t of coal equal to 2.70 t of firewood and to 1.39 t of charcoal.[9] Since charcoal consumption was lower than firewood and its relative price fluctuated at a slightly higher level, it is sufficient for our purposes to compare the price

Table 1.10 Coal consumption in Nagano Prefecture and Hirano Village (in tons)

Year	Nagano Prefecture (a)	Hirano Village (b)	(b)/(a) (%)
1890	143	n.a.	–
1891	368	200	54.3
1892	1,428	854	59.8
1893	2,893	1,716	59.3
1894	1,490	1,149	77.1
1895	3,166	2,290	72.3
1896	14,451	13,149	91.0
1897	15,615	n.a.	–
1898	14,227	11,949	84.0
1899	18,271	12,000	65.7
1900	10,606	n.a.	–
1901	6,406	n.a.	–
1902	6,639	n.a.	–
1903	22,005	n.a.	–
1904	27,700	n.a.	–

Sources: 1890–1893, Naganoken Kangyōka (NKN), nos 13, 14, 16; 1894–1904, Naganoken Tōkeisho (NT) (1883), Hiranomura Yakuba (HYa), (HYb), (HYc), (HYe)
Notes: Figures for Hirano in 1899 in the original source were probably misprinted and have therefore been revised. All coal was consumed in private factories

---

[9]Firewood had 51.4 % of the heat efficiency of charcoal, and 37.0 % compared to Hokkaidō coal (Tetsudōshō 1925; Eguchi and Hidaka 1937, vol. 3).

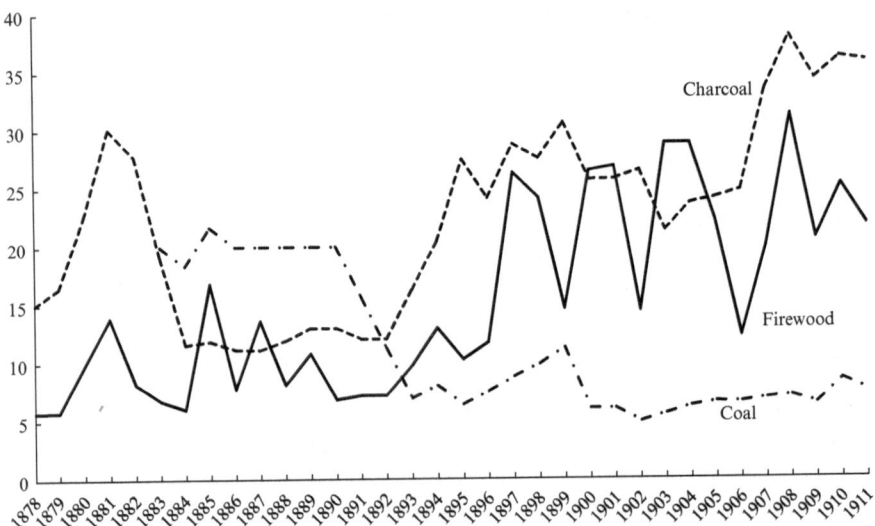

Sources: Prices of firewood and charcoal for 1878-1884, Eguchi and Hidaka (1937), vol.3, pp. 1282-83; for 1885-1896, Naganoken Kangyōka (NKN), nos 8-19; for 1897-1911, Naganoken (NT). Prices of coal for 1883-1890, Naganoken (NT); 1893, Naganoken Kangyōka (NKN), no.16; 1894-1911, Naganoken (NT).

Fig. 1.2 Relative prices of firewood, charcoal and coal in Nagano, 1878–1911 (in yen)

fluctuations of firewood and coal. Firewood prices fluctuated over a wide range. From the late 1880s to the early 1890s, Hirano relied on both the transfer of trees from the neighbouring state and imperial forest land, and increasing deliveries from neighbouring districts. This prevented rises in price, but after 1897 the price moved at a relatively high level with wider fluctuations. In contrast, coal prices decreased rapidly in the early 1890s due to the expansion of the domestic coal market that was being facilitated by the expansion of the railway network. The conclusion can therefore be drawn that the increase in coal consumption after 1895 resulted not simply from the high price of firewood, but from a combination of the relatively low coal price in the early 1890s and the increasing price of firewood after 1897.

However, the relative decline in coal prices and the consequent short-term increase in coal consumption in the early 1890s did not mean that a permanent shift had taken place from firewood to coal. The completion of the Shin'etsu railway facilitated the transport of light-weight goods such as cocoons from Ōya along the railway to Suwa via the main road of Nakasendō, but bulky goods such as coal were still difficult to move. In addition, since the technological problems in using coal mentioned before had not been resolved, it could only be used for around 30 % of all basins during the decade from the late 1890s (Hiranomura 1932, vol. 2). The relative decline in the use of power boilers from the late 1880s to the early 1890s, and the rise in the combined use of water power and steam power shown in Table 1.5, were in part stimulated by these developments. In the case of Hirano, steam power increased again in 1905, but in Suwa as a whole, only 36 % of factories used steam power, while 58 % still relied on water power.

In 1903, the Shinonoi railway line was extended from Matsumoto to Shiojiri, and the cable railway between Shiojiri and Hirano began operations. This facilitated local transportation of coal, resulting in increased consumption. The extension of the Chūō railway line to Okaya in 1905 allowed the transportation of coal from the more distant Jōban and Kyūshū areas. When the Chūō railway line came into full operation in 1911, 60 % of the total 55,000 t of coal that arrived at Okaya station came from Jōban and 15 % from Kyūshū (Tetsudōin 1916, vols 1 and 2; Eguchi and Hidaka 1937, vol. 3).

Filatures in Hirano in 1906 consumed 240 t of firewood and 139 t of charcoal, but coal was clearly the main source of fuel, at 10,500 t. The three major filatures of Kaimeisha, Oguchigumi and Shin'eisha accounted for 61 % of Hirano's total coal consumption (HYe). In other words, there was a clear contrast between large consumers such as Kaimeisha, Oguchigumi and Shin'eisha that made a rapid shift to coal, and those filatures that continued to depend on firewood and charcoal. In Suwa as a whole, as we have seen, there were variations in the shift to coal, and dependence on water power continued. A combination of factors account for this contrast: a lack of skilled boilermen, difficulties in the transport of coal, the dramatic rise in firewood prices of the late 1890s, and the availability of cheap water power from the Tenryū river. These factors caused silk producers either to make the temporary investment of funds needed to switch to combined steam boilers that used coal for both heating and power generation, or to find a short-term solution to their fuel problems by maximizing their exploitation of traditional water power.

Additional research into individual filatures is needed before we can judge which alternative held the comparative advantage, but we assume that those with financial resources who also experienced relative difficulty in obtaining access to water power chose the former, while those located near Tenryū river and/or without available surplus funds tried to solve their immediate fuel problems by choosing the latter.[10] This caused a temporary increase in the use of water power in Suwa, and also led to continuing dependence on firewood and charcoal. As a consequence, there was no sudden decline in either the demand for firewood or its price. The decisive shift from firewood to coal in silk reeling did not take place until after the Russo-Japanese war of 1904–1905, when improved coal-using steam boilers became available at cheaper prices. Electric power became more common after the establishment of the Suwa Electric Co. in Shimo-Suwa in 1897; by the time of World War I, the industrial power market in this region was dominated by electricity (Suwa Denki Kabushiki Kaisha 1896a, b, 1919).

---

[10]Since water-powered factories and steam/water powered factories had about the same productivity per basin, it can be said that the former were superior in terms of production costs (Takamura 1995).

5   Conclusion

This article has examined the link between deforestation and the development of the silk reeling industry in the region of Suwa in Nagano prefecture from the 1870s to the 1900s, and the subsequent shift from firewood to coal. Since the late Tokugawa period, firewood for the silk reeling industry had come from land held in common by neighbouring villages. From the 1870s, the development of silk reeling as Japan's most important export industry encouraged a shortage of firewood. In a classic illustration of 'the tragedy of the commons', the main victims of the increased demand were trees and grass growing on common lands (much of which was mountainous), since there were only weak restrictions on use. By the mid-1880s, traditional sources of firewood were being supplemented by the transfer to silk producers of trees from state and imperial forests. The Nagano prefectural government tried to encourage tree-planting, without success. In the early 1890s it became necessary to transport firewood from neighbouring districts.

The steam boilers that were adopted by the silk reeling industry from the late 1870s were cheap to buy, but too weak for use with coal. As a result, coal did not become dominant until around the turn of the twentieth century, when steam boilers were more robust and the gap in the relative prices of coal and firewood had been reduced. However, it is true that the increasing firewood shortage had already sparked off the shift to coal as an alternative fuel in the early 1890s. The decline of the relative price of coal resulted from the increase in coal production and the expansion of the railway distribution network. Both these factors facilitated the rapid change from firewood to coal as the main source of industrial energy.

It is likely that the fuel crisis was one factor behind the trend for raw silk producers to extend their activities beyond Nagano prefecture in the late 1890s, alongside the chance to obtain better quality cocoons (Hiranomura 1932, vol. 2). The Nagano silk reeling industry may not be typical since, with the notable exception of Hirano, extensive use of water power was possible. Even so, it is significant because it shows us how the transformation of a traditional industry such as silk reeling into a leading exporter could not occur without causing lasting damage to the local environment.

Primary Sources

Hiranomura Yakuba (HYa) *Sangyō Kankei Shorui* (Documents related to Local Industries), 1, 1875–96 (Okaya Silk Museum, 5-91-1).

Hiranomura Yakuba (HYb) *Sangyō Kankei Shorui*, 2, 1897–1906 (Okaya Silk Museum, 5-91-2).

Hiranomura Yakuba (HYc) *Sangyō Kankei Shorui*, 6, 1897–1906 (Okaya Silk Museum, 5-95).

Hiranomura Yakuba (HYd) *Seishi Kankei Shorui* (Documents related to Local Silk Reeling), 1, 1875–1882 (Okaya Silk Museum, 5–108).

Hiranomura Yakuba (HYe) *Seishi Kankei Shorui*, 15, Miscellaneous (Okaya Silk Museum, 5–119).

Hiranomura Yakuba (HYf) *Yokokawayama Kankei Shorui* (Documents related to Mt Yokokawa), Date unknown (Okaya Silk Museum, 1–347).

Hiranomura Yakuba. 1906. *Yokokawayama Shokurin Ichiranhyō* (Table of Tree Planting on Mt Yokokawa), June 1906. In: Hiranomura Yakuba (HYf).
Hiranomura and Kawagishimura Kangyō Kakari. 1885. *Shokurin Daichō* (Ledgers related to Tree Planting), December 1885 (Okaya Silk Museum, 1–338).

References

Andō, Seiichi. 1992. *Kinsei kōgaishi no kenkyū* (Studies in the history of pollution in pre-modern Japan). Tokyo: Yoshikawa Kōbunkan.
Chiba, Tokuji. 1991. *Hageyama no kenkyū* (Studies in mountain deforestation), revised and expanded edn. Tokyo: Soshiete.
Chiba, Tokuji. 1953. Chūbu Shinano no tokuchochi (Deforestation in central Shinano). *Chirigaku Hyōron* 26(7).
Crosby, Alfred W. 1986. *Ecological imperialism: The biological expansion of Europe, 900–1900.* Cambridge: Cambridge University Press.
Eguchi, Zenji, and Hidaka, Yasoshichi (eds.). 1937. *Shinano sanshigyō-shi* (A history of the Shinano silk industry), 3 vols. Nagano: Dai-Nihon Sanshikai Shinano Shikai.
Elvin, Mark, and Liu, Ts'ui-jung (eds.). 1998. *Sediments of time: Environment and society in Chinese history.* Cambridge: Cambridge University Press.
Fukushima, Masao, et al. (eds.). 1956. *Meiji nijūroku-nen zenkoku sanrin gen'ya iriai kankō chōsa shiryō* (Documents from the 1893 national survey of the customs related to commonly held forests and fields), 1. Tokyo: Shinrin Shoyūken Kenkyūkai.
Fukushima, Masao, et al. (eds.). 1958. *Shōwa go-nen zenkoku sanrin gen'ya iriai kankō chōsa shiryō* (Documents from the 1930 national survey of the customs related to commonly held forests and fields), 2, Nagano Prefecture. Tokyo: Shinrin Shoyūken Kenkyukai.
Hayami, Yujirō. 1995. *Kaihatsu keizaigaku* (Development economics). Tokyo: Sōbunsha.
Hirano, Yasushi. 1990. *Kindai yōsangyō no hatten to kumiai seishi* (Silk reeling associations and the development of sericulture in modern Japan). Tokyo: Tokyo Daigaku Shuppankai.
Hiranomura. 1932. *Hirano sonshi* (A topography of Hirano Village), 2 vols. Nagano: Hiranomura Yakuba.
Horie, Sangorō. 1930. *Suwako hanran sanbyakunen-shi* (Three hundred years of flooding by Lake Suwa). Kami-Suwa: Suwako Hanranshi Kankōkai.
Ishii, Kanji. 1972. *Nihon sanshigyō-shi bunseki* (A historical analysis of the modern Japanese silk industry). Tokyo: Tokyo Daigaku Shuppankai.
Kamioka, Namiko. 1984. *Nihon no kōgai-shi* (A history of pollution in Japan). Tokyo: Sekai Shoin.
Kitajima, Masamoto (ed.). 1970. *Seishigyō no tenkai to kōzō* (The development and structure of the silk reeling industry of Japan). Tokyo: Hanawa Shobō.
Kiyokawa, Yukihiko. 1995. *Nihon no keizai hatten to gijutsu fukyū* (The diffusion of technology and the economic development of Japan). Tokyo: Tōyō Keizai Shinpōsha.
Makino, Fumio. 1996. *Manekareta Prometheus* (Invited Prometheus). Tokyo: Fūkōsha.
Matsumura, Satoshi. 1992. *Senkanki Nihon sanshigyō-shi kenkyū* (Studies in the history of the Japanese silk industry during the interwar period). Tokyo: Tokyo Daigaku Shuppankai.
Matsunami, Hidezane. 1919. *Meiji ringyō shiyō* (A history of Meiji forestry). Tokyo: Dai-Nihon Sanrinkai.
Naganoken Kangyōka (Nagano Prefecture, Division of the Promotion of Industry) (NKG). 1885. *Kangyō geppō* (Monthly report on the promotion of industry) Nagano: Naganoken.
Naganoken Kangyōka (NKN). 1878–1896. *Naganoken kangyō nenpō* (Annual report on the promotion of industry in Nagano Prefecture). Nagano: Naganoken.

Naganoken Sanshigyō Kumiai. 1887. *Naganoken sanshigyō kumiai torishimarijo nenpō* (Annual report of the regulatory agency of silk reeling industry associations in Nagano prefecture), 1, December.

Naganoken Sanshigyō Kumiai. 1888. *Naganoken sanshigyō kumiai torishimarijo nenpō*, 2, December.

Naganoken-shi Kankōkai (ed.). 1980. *Naganoken-shi: kindai shiryō-hen* (The history of Nagano Prefecture, documents related to the modern period), 5–3. Nagano: Naganoken-shi Kankōkai.

Naganoken-shi Kankōkai (ed.). 1986. *Naganoken-shi: Kindai shiryō-hen.* 5–4.

Naganoken Suwagun Yakusho. 1905–1918. *Suwagun gunchi ippan* (A general survey of the Suwa region). 1–6.

Naganoken Tōkeisho (NT). 1883–1911. *Naganoken tōkeisho* (Annual statistical survey of Nagano Prefecture). Nagano: Naganoken.

Nakamura, Kichiji (ed.). 1956. *Sonraku kōzō no shiteki bunseki* (A historical analysis of village structure). Tokyo: Nihon Hyōronsha.

Nakamura, Kichiji, et al. 1962. *Hōkenki kaitai nōson no kenkyū* (Studies on rural communities during the decline of the feudal system in Japan). Tokyo: Sōbunsha.

Nōshōmushō Nōmukyoku (Ministry of Agriculture and Commerce, Division of Agriculture). 1895, 1898, 1902, 1907. *Zenkoku seishi kōjō chōsahyō* (National survey of silk filatures). 1 (October 1895), 2 (April 1898), 3 (March 1902), 4 (June 1907).

Oda, Yasunori. 1983. *Kindai Nihon no kōgai mondai* (Problems of pollution in modern Japan). Kyoto: Sekai Shisōsha.

Ōi, Takao. 1973a. Kindai ni okeru ringyō hatten no kisoteki zentei (Basic prerequisites for the development of forestry in modern Japan) (1). *Shinano* 25(11).

Ōi, Takao. 1973b. Kindai ni okeru ringyō hatten no kisoteki zentei (Basic prerequisites for the development of forestry in modern Japan) (2). *Shinano* 25(12).

Ōishi, Kaichirō. 1968. Nihon seishigyō chinrōdō no kōzōteki tokushitsu (Structural characteristics of wage labour in the Japanese silk reeling industry). In *Kokumin keizai no shoruikei* (Patterns of national economies), ed. Takeyoshi Kawashima, and Tomoo Matsuda. Tokyo: Iwanami Shoten.

Okayashi. 1976. *Okayashi-shi* (The history of the city of Okaya), vol. 2. Okaya: Okayashi.

Pomeranz, Kenneth. 2000. *The great divergence: China, Europe, and the making of the modern world economy*. Princeton: Princeton University Press.

Ponting, Clive. 1992. *A green history of the world: The environment and the collapse of great civilizations*. New York: St. Martin's Press.

Shimo-Suwa Chōshi Hensan Iinkai (ed.). 1963. *Shimo-Suwa chōshi* (A topography of the town of Shimo-Suwa), vol. 1. Tokyo: Kōyō Shobō.

Shimo-Suwa Chōshi Hensan Iinkai (ed.). 1969. *Shimo-Suwa chōshi*, vol. 2. Tokyo: Kōyō Shobō.

Shinano Sanrinkai. 1902–1928. *Shinano Sanrin kaihō* (Bulletin of the Shinano Forest Association), 1 (November 1902) – 39 (April 1928).

Shinano Sanrinkai (ed.). 1904. *Naganoken shinrin tōkeisho* (A statistical survey of forests in Nagano Prefecture), 1.

Sugiyama, S. 1988. *Japan's industrialization in the world economy, 1859–1899: Export trade and overseas competition*. London: Athlone Press.

Suwa Denki Kabushiki Kaisha (Suwa Electric Power Co.). 1896a. *Setsuritsu mokuromisho* (Company prospectus).

Suwa Denki Kabushiki Kaisha. 1896b. *Sōritsu shutsugansho* (Application to found the company).

Suwa Denki Kabushiki Kaisha. 1919. *Sōgyō nijū-shūnen kinen* (Twenty-year anniversary of the company).

Suzuki, Jun. 1996. *Meiji no kikai kōgyō* (The machine industry of Meiji Japan). Kyoto: Minerva Shobō.

Takamura, Naosuke. 1995. *Saihakken Meiji no keizai* (The Meiji economy rediscovered). Tokyo: Hanawa Shobō.

Tetsudōin (Railway Agency). 1916. *Honpō tetsudō no shakai oyobi keizai ni oyoboseru eikyō* (The socio-economic effects of railways in Japan), 3 vols. Tokyo: Tetsudōin.

Tetsudōshō (Ministry of Railways) 1925. *Mokuzai ni kansuru keizai chōsa* (An economic survey of timber). Tokyo: Tetsudōshō.

Tōjō, Yukihiko. 1990. *Seishi dōmei no jokō tōroku seido* (Silk associations and their systems for registering female workers). Tokyo: Tokyo Daigaku Shuppankai.

Totman, Conrad. 1989. *The green archipelago: Forestry in pre-industrial Japan*. Berkeley/Los Angels: University of California Press.

Tsutsui, Michio. 1978. *Nihon rinsei-shi kenkyū josetsu* (An preliminary study of the history of forest policies in modern Japan). Tokyo: Tokyo Daigaku Shuppankai.

Umemura, Mataji, et al. (eds.). 1966. *Chōki keizai tōkei* (Long-term economic statistics), 9, Agriculture and forestry. Tokyo: Tōyō Keizai Shinpōsha.

Worster, Donald. 1993. *The wealth of nature: Environmental history and the ecological imagination*. Oxford: Oxford University Press.

Wrigley, E.A. 1988. *Continuity, chance and change: The character of the industrial revolution in England*. Cambridge: Cambridge University Press.

Yagi, Haruo. 1960. *Nihon kindai seishigyō no seiritsu* (The formation of the silk reeling industry in modern Japan). Tokyo: Ochanomizu Shobō.

Yamaguchi, Kazuo (ed.). 1966. *Nihon sangyō kin'yū-shi kenkyū: Seishi kin'yu-hen* (Studies in the history of industrial financing in Japan: The silk reeling industry). Tokyo: Tokyo Daigaku Shuppankai.

Yano, Makoto (ed.). 2008. *The Japanese economy: A market quality perspective*. Tokyo: Keio University Press.

Chapter 2
The Government Railways and the Procurement of Railway Sleepers in Prewar Japan

Asuka Yamaguchi

Abstract In the Japanese industrialization process, it was essential for industries to procure stable supplies of timber because of its vital role as a raw material, as a source of energy, and for use in construction. The purpose of this article is therefore to examine the issue of timber procurement through a case study of the supply of timber for use as rail sleepers by the Japanese Government Railways (JGR) in the prewar period. JGR's need for wooden sleepers increased as its network expanded, and in 1909 it switched from competitive tendering to sole-source contracts in order to secure stable supplies at prices within the annual budget. In 1930 JGR changed to invited tendering in order to secure low prices at a time of budget restrictions, but in 1933 economic conditions improved and it returned to sole-source contracts. Thus JGR adapted its supply methods to meet changes in both its budget and in the timber market. However, despite efforts to utilize a wider range of trees and preservative treatments, it experienced increasing difficulties in finding supplies from the late 1930s.

Keywords Wood • Timber • Sleeper • Railway • Environment

1 Introduction

In the Japanese industrialization process, wood was widely used by modern and traditional industries as well as in the residential sector as an energy source, a raw material and a construction material. In the seventeenth century, or the early Tokugawa period, logging was conducted and wood was used mainly for

---

This is a translation of an article that originally appeared in *Shakai Keizai Shigaku* 76(4) (February 2011), pp. 49–72.

A. Yamaguchi (✉)
Graduate School of Economics, Nagoya City University, 1 Yamanohata, Mizuho-cho, Mizuho-ku, Nagoya 467-8501, Japan
e-mail: yamaguchi@econ.nagoya-cu.ac.jp

constructing castle towns and for developing arable land in response to population increases. Beginning in the nineteenth century, however, industrial development changed the supply and demand structure for wood. This was especially apparent in the modern period when wood demand accelerated with advancing industrialization; not only did the demand for firewood by traditional industries increase, but demand for wooden materials also emerged in modern industries. In such conditions, it was a challenge for industries to secure stable supplies of wood over the long term.

In research on the history of wood, a number of previous studies examined specifically the history of forestry, unions, and village communities. The body of literature increases enormously if we include prefectural and municipal history. These studies mainly consider the history at a regional level and examine the process of wood production, which includes afforestation, logging, processing, the use of water transportation techniques, and use of forests by actors like forestry managers, unions, and regional entities. Thus, past research on the history of wood involved little analysis of the distribution and consumption of wood. In other words, despite the fact that wood played an important role in the industrialization process, research on the history of wood use in Japan has only briefly discussed its role in the industrialization process.

In the field of economic history, wood has been examined as an energy source (firewood and charcoal), and energy consumption trends in the residential sector and the traditional industries have been analyzed from the standpoint of the development and the selection of technologies (Makino 1996; Taniguchi 1998). In the modern and subsequent periods, demand for firewood and charcoal was large because these resources were continuously used in the residential sector and by traditional industries. However, wood played another important role in the industrialization process. Wood was used as construction and raw materials in modern industries such as the railway, coal mining, and paper industries. Although the share of such timber use in total wood consumption was not necessarily large, demand for timber grew rapidly. Indeed, the generation and expansion of timber demand in modern industries significantly changed the supply and demand structure of the Japanese wood market. In order to better understand the role of timber in the industrialization process, therefore, it is necessary to analyze the market for timber that was used in construction and as a raw material. To this end, it is also useful to examine changes in the market for firewood and charcoal and in energy use. Moreover, past research on economic history focused mainly on procurement and use of raw materials and barely examined construction materials, despite the fact that they are essential factors that define the activities of industries. A previous paper by the author shed light on the procurement and use of timber in the coal-mining industry (Yamaguchi 2008, 2009), but comprehensive analysis of the procurement and use of construction materials has yet to be undertaken.

Against this background, this paper considers wooden sleepers in the railway industry as an example of the use of timber as an industrial construction material. Like the mining industry, the railway industry was a central target for government support under the industrial promotion policy of the Meiji government, and approximately 50 % of the Ministry of Industry's total expenditure for industrial promotion from December 1870 (when the Ministry was established) through December 1885 (when it was abolished) was channeled to the railway industry (Kobayashi 1977). In the modern and subsequent periods, the railway industry played an important role in Japan's industrialization, particularly as a growth engine driving the rise of enterprises during the 1880s and also as the main transportation industry supporting the expansion of domestic markets. Earlier research on the railway industry has mainly focused on railway policy, financing, railway technology, functions of the railway as a transportation system, and effects on regional economies (Oikawa 1992; Nakamura 1998; Matsushita 2004). In regard to railway materials, there is only one study on vehicles by Sawai Minoru (Sawai 1998). Based mainly on statistics and investigation documents from the Japanese Government Railways (JGR),[1] documents in the Railway Museum, and relevant documents from the sleeper dealer Hasegawa Shōten (Hasegawa Timber Co.), this paper sheds light on the use of timber as an industrial construction material by examining the procurement and use of sleepers by JGR.

2 The Position of Sleepers in Timber Consumption

2.1 Consumption of Wood

Wood is mainly used as a fuel, either firewood or charcoal, and as a material for building construction, mine pillars, rail sleepers, electric poles, pulpwood, and military installations. In the early modern and subsequent periods, firewood and charcoal were used in traditional industries such the salt, ceramic, and silk reeling industries. In the 1880s and 1890s, the share of fuel in total wood consumption was more than 80 % (36 million cubic meters). In the early twentieth century, energy sources of various industries shifted from firewood and charcoal to coal, electricity, and petroleum; however, fuelwood consumption was consistently above 28 million cubic meters because households and traditional industries continued to use firewood and charcoal. The consumption of timber was around 6 million cubic meters in the 1880s, and it started rising along with industrial development during the economic boom which followed the deflationary fiscal policy by Finance Minister, Matsukata Masayoshi in the early 1880s. Subsequently, consumption reached 11 million cubic meters at the beginning of the twentieth century and 17 million cubic meters in the period of the economic boom during World War I. The consumption of timber trended upward after 1920 except in the periods of postwar recession (1920–22) and depression (1928–32), and exceeded 28 million cubic meters starting in the second half of the 1930s due to a rapid rise in the demand

---

[1]The government agency in charge of operating the national railways continually changed. This paper uses the generic term "Japanese Government Railways" in referring to the national railway agencies.

for military materials. As these figures show, the structure of the demand for timber changed significantly with the rise in industrial development in the modern period.

With regard to the consumption of timber in different sectors, the construction sector had the largest share which was around 50 % until the 1900s. However, starting in the 1910s, the share fell to between 20 % and 40 % as timber consumption increased in other industries. The second largest share was held by the mining sector, at 1–5 % in the 1880s and 1890s and consistently at around 10 % early in the twentieth century. The consumption of pulpwood used as a raw material for paper rose in the twentieth century as more demand for paper was generated by an increased volume of newspapers and magazines issued and by the start of the system of government-designated textbooks; its share was 3–6 % after the second half of the 1910s. Also, with development of the transportation infrastructure being promoted, timber consumption in public projects such as the construction and the repair of roads and ports increased, and its share was 2–6 % from the early 1880s to 1940 (Umemura et al. 1966).

The share of sleepers in timber consumption was 1–2 % consistently from the end of the 1880s onwards. This share was small compared to the kinds of timber consumption mentioned above; however, this does not necessarily mean that the procurement of timber for railroad sleepers was easy. This is because tree varieties as well as the size and shape of timber which is used for different purposes vary. Based on the amount of trees felled for timber during 1905 through 1921 for different tree varieties, it is estimated that the share of pine and cedar in the timber market in Japan was 50–70 %, and that the share of chestnut timber, which was in high demand for sleepers in Honshū, was merely 2–3 %. In addition, tree varieties that are suitable for sleepers, such as chestnut, hiba, and hinoki (Japanese cypress), were also used for construction materials, furniture, firewood, and charcoal. Therefore, timber for sleepers was in a competitive relationship with timber used for such purposes. Also, since high-quality, large-diameter logs that are not bent or cracked are needed for sleepers, the market availability of timber for sleepers was limited. Timber for sleepers mainly comes from domestic sources. According to statistics from the Ministry of Agriculture and Commerce for 1905–1914, the sleeper-timber-producing regions are Hokkaidō (approximately 40–60 %, including exports), the Tōhoku region, including Aomori, Iwate, and Fukushima (approximately 20 %), the Chūbu region, including Aichi, Nagano, and Gifu (approximately 4 %), the Kinki region, including Kyōto and Hyōgo (approximately 4 %), the Chūgoku region, including Hiroshima and Shimane (approximately 8 %), and the Kyūshū region, including Ōita and Kagoshima (approximately 2 %) (Nōshōmushō 1907–1921) (Fig. 2.1).[2]

---

[2]Sleepers from Hokkaidō (fraxinus spaethiana, kalopanax septemlobus, katsura, quercus serrata, etc.) were different from sleepers from the main island of Japan (chestnut, hiba, hinoki, etc.), in terms of tree varieties used.

Region	No.	Prefecture	Region	No.	Prefecture
Hokkaidō	1	Hokkaidō		24	Mie
	2	Aomori		25	Shiga
	3	Iwate		26	Kyōto
Tōhoku	4	Miyagi	Kinki	27	Ōsaka
	5	Akita		28	Hyōgo
	6	Yamagata		29	Nara
	7	Fukushima		30	Wakayama
	8	Ibaraki		31	Tottori
	9	Tochigi		32	Shimane
	10	Gunma	Cyūgoku	33	Okayama
Kantō	11	Saitama		34	Hiroshima
	12	Chiba		35	Yamaguchi
	13	Tōkyō		36	Tokushima
	14	Kanagawa	Shikoku	37	Kagawa
	15	Niigata		38	Ehime
	16	Toyama		39	Kōchi
	17	Ishikawa		40	Fukuoka
	18	Fukui		41	Saga
Cyūbu	19	Yamanashi		42	Nagasaki
	20	Nagano	Kyūsyū	43	Kumamoto
	21	Gifu		44	Ōita
	22	Shizuoka		45	Miyazaki
	23	Aichi		46	Kagoshima

※ Honsyū is the region excluding Hokkaidō.

Fig. 2.1 Map. Prefectures and major place names

2.2 Estimated Consumption of Sleepers

The consumption of sleepers nationwide and by JGR in particular, increased in the prewar period. Demand for sleepers increased year by year with the extension of railways not only because sleepers were used in the construction of new railway lines or additional tracks, but also because sleepers needed to be replaced due to decay or damage. The volume of sleepers consumed nationwide and by JGR shown in Fig. 2.2 is calculated by multiplying the distance (single-track equivalent, i.e., the sum length of the sidetracks in a multiple-track line added to the length of the main track) by the number of sleepers used per kilometer in new construction and improvement/repair. The number of sleepers used per kilometer varies depending on the time period, the type of the site (e.g., level ground, bridge, tight curve, and steep slope), and the specifications used by railway companies, but this number is assumed to be 1,250–1,500 in new construction and 130–150 in improvement/repair, based on the documents available. As shown in Fig. 2.2, sleeper consumption rapidly increased after 1906 and 1907 (when railway operation was nationalized),

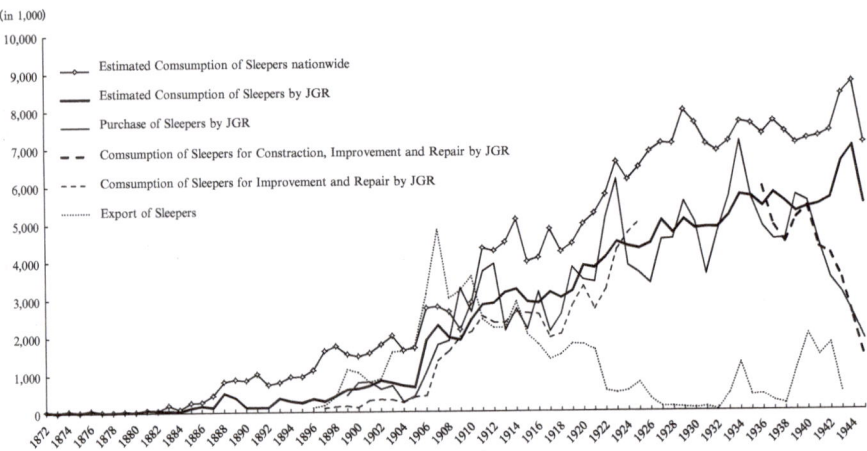

Sources: Nihon Tōkei Kyōkai (2006), pp.506-508; Naikaku Tōkeikyoku (1896-1937); Ōkurashō (1902-1943); Nihon Kokuyū Tetsudō (1971), p.645; Tetsudōsagyōkyoku (1898-1905); Tetsudōchō (1906 and 1907); Tetsudōin (1908-1920); Tetsudōin (1916-1919); Tetsudōshō (1920-1941); Yoshitsugu (1951), p.340; Nihon Makuragi Kyōkai (1959), p.2.

Note: The amount of sleepers consumed is calculated by multiplying the distance of railways by the number of sleepers used per kilometer in new construction and improvement/repair. The number of sleepers used per kilometer is assumed to be 1,250 (1872-1880), 1,380 (1881-1910), 1,500 (1911-1945) in new construction, and 130 (1872-1880), 140 (1881-1910), 170 (1911-1945) in improvements/repairs.

Fig. 2.2 The amount of sleepers consumed nationwide and by JGR, 1872–1945

tended to flatten during World War I, started to rise again in the 1920s, and tended to flatten again thereafter (from the early 1930s onwards in the case of JGR).

After the railway between Shinbashi and Yokohama opened in 1872, major cities and ports were connected by railways, and a nationwide arterial network was mostly completed in the first half of the 1900s. After the Russo-Japanese War of 1904–05, mention began to be made that improvement in the international balance of payments and growth in exports to the Chinese market would require lower railway fares and the development of an integrated system of transportation to Korea and Manchuria. At the same time, a movement toward railway nationalization, which was led by the military advocating the importance of wartime military transportation, gained strength. As a result, 17 private railway companies were nationalized in 1906 and 1907. The proportion of JGR's operating kilometers to operating kilometers nationwide increased from 29 % (2,413 km) at the end of 1905–82 % (7,152 km) at the end of 1907, and JGR's share in the sleeper market rose to more than 80 %. In the 1920s, electric railways and subways in cities like Tokyo and Ōsaka as well as regional railways were rapidly developed. Consequently, the proportion of JGR's operating kilometers dropped to 67 % by 1940. However, in terms of total distance, which takes into account the length of multiple-track lines, JGR's proportion still exceeded 70 % in 1940. Moreover, in terms of freight (kilometers per ton) and passengers (kilometers per passenger) which affect the wear rate of sleepers, JGR's share was 91–99 % and 82–94 %, respectively. This puts JGR's share in the tie market at a minimum of 70 % after railway nationalization.

3 The Period Before Railway Nationalization of 1906

3.1 Procurement of Sleepers Through the Sale of Government Forests

In building a railway between Shinbashi and Yokohama, the government created the Railway Office (*Tetsudōgakari*) in the Ministry of Public Affairs and Finance (Minbu-Ōkurashō).[3] The construction work was conducted under the instruction of British chief engineer Edmund Morel. Most of the materials used in the construction were imported from Britain, and expensive steel sleepers were imported. Thus, railway construction expenditure was large not only for the materials purchased from foreign countries but also for the foreign advisers hired by the Japanese government (Nakamura 1998). Following the recommendations from Morel that Japanese wooden sleepers were cheaper than steel sleepers and were more suitable to the environment of Japan, the government decided to limit the purchase of steel sleepers as well as to use domestic timber for subsequent railway construction (Nihon Kokuyū Tetsudō 1969).

In August 1872, the Administrative Rules of *Tetsudōryō* were enacted, and *Tetsudōryō* of the Ministry of Industry (*Kōbushō*) was required to obtain timber, earth, and rocks needed for railway construction in a flexible manner through consultation with relevant authorities. In accordance with the rules, *Tetsudōryō* procured timber for sleepers, girders, train vehicles, and train stations, as well as for foundation work for embankments and bridges, through the Ministry of Finance (Ōkurashō) which then had jurisdiction over the government forests. These forests formerly belonged to the Tokugawa Shogunate, feudal clans, temples, and shrines, but became the property of the Meiji government due to the return of the lands and people from the feudal lords to the emperor in 1869 (*Hanseki Hōkan*), the abolition of the feudal system, and the establishment of prefectures in 1871 (*Haihan Chiken*). For example, in the first railway construction between Shinbashi and Yokohama, *Tetsudōryō* procured the necessary timber from timber being stored at Fukagawa-Sarue which was under the Ministry of Finance's jurisdiction. In the construction of the railway between Kyōto and Kōbe, the Great Council of State (*Dajōkan*) notified Ōsaka, Kyōto, and neighboring prefectures that it would sell to merchants, through a tendering process, the trees (with the land) in the government forests in areas that could be logged, and that it would purchase processed products from these trees. Thus, the Ministry of Industry's *Tetsudōryō* procured timber from Aichi, Wakayama, Mie, Kagawa, and Yamaguchi through the Finance Minister. However, it was difficult to secure high-quality, large-diameter logs. Also, investigations for and the procedure related to securing timber took substantial time. Therefore, Masaru Inoue, the head of *Tetsudōryō*, requested through the Minister of Industry

---

[3]In what follows, *Tetsudōryō*, *Tetsudōkyoku*, *Tetsudōchō*, and *Tetsudōsagyōkyoku* are all names of the bureau in charge of railway operation at a given time.

the direct sale of government forests to the Ministry of Home Affairs (*Naimushō*), and the jurisdiction over the government forests was transferred from the Ministry of Finance to the Ministry of Home Affairs in January 1874. This change made it possible for officials from the Ministry of Engineering and Industry to purchase timber through direct contact with prefectural governments (Nihon Kokuyū Tetsudō 1969). In the construction of the railway between Ueno and Morioka (opened in 1891), six people from Kitakunohegun, Iwate, and one person from Sannohegun, Aomori, submitted a joint petition to the Ministry of Industry and Engineering's *Tetsudoryō* for a purchase of 300,000 chestnut sleepers (priced at 112,500 yen) (Karumichō 1975). Although this seems to be a case of a direct purchase by timber dealers, in the 1870s and the 1880s sleepers were mainly procured through the sale of government-owned forests in various prefectures.

3.2 Procurement of Sleepers Through Competitive Tendering

The Public Accounting Act promulgated in February 1889 stipulated that government procurement be conducted in principle through competitive tendering. The accounting regulations enacted in May of the same year required that those who were eligible to participate in competitive tendering and to sign a contract must have engaged in the relevant business for two or more years, and that a payment of 5 % or more of the quoted price (if participating in the tendering process) and 10 % or more of the security deposit (if signing a contract) must be made in cash or public bonds. Information on the tendering contest, such as the time and place and the amount of security deposit required, was posted or announced in newspapers or government bulletins. Based on the Public Accounting Act and the accounting regulations, the Home Ministry's *Tetsudōchō* enacted regulations on the purchase and sales conducted by the agency in October 1890 and started to procure sleepers through competitive tendering. Complying with the regulations, the Home Ministry's *Tetsudōchō*, for instance, announced in the government gazette for February 4, 1891 that those who wished to participate in a tendering process for hinoki, hiba, and chestnut sleepers (30,000 sleepers for each kind) should mail their tender form to the agency by February 25 of the same year. Responding to the announcement, the participants mailed their tender form containing their quote (unit price and total amount) by the specified date (Dajōkan Monjokyoku 1891).

Let us now look at the sleeper dealers who participated in competitive tendering and the content of the contracts. Table 2.1 shows petitions for cancellation of delivery contracts for sleepers submitted from 1893 through 1902 to the Ministry of Communications' (*Teishinshō*) *Tetsudōkyoku* (the Ministry of Communications' *Tetsudōsagyōkyoku* from August 1897). As the table shows, sleeper dealers came from Hokkaidō and the Tōhoku, Chūbu, Kinki, and Chūgoku regions, but not from Kyūshū or Shikoku. The listed contracts are considered to have been for procuring sleepers used in the construction of the railways between Fukushima and Aomori and between Tsuruga and Toyama, which was intensified from 1896,

Table 2.1 Petitions for cancellation of delivery contracts for sleepers submitted to the Ministry of Communications' Tetsudōkyōku (Tetudōsagyōkyoku), 1893–1902

No.	Sleeper dealer	Address	Date of contact	Deadline for completion delivery	Tree varieties	Price of sleepers contracted (in yen)	Number of sleepers contracted	Number of sleepers delivered	Sleeper producing regions	Sleeper delivery destination	Petitions for cancellation of delivery contracts — Date of submission	Reason
1	Hanshichi Gotō	Hokkaidō	22 Aug. 1899				6,400	20,000	10,000			1 Dec. 1899	
	Daigorō Sekikawa (Hanshichi Gotō's agent)	Hokkaidō		Dec. 1899	Class 2			5,000	2,500		Noshiro	1 Dec. 1899	Transport difficulties in bad weather
1'	Hanshichi Gotō	Hokkaidō		31 June 1898				20,000					
	Yohachi Hirata (Hanshichi Gotō's agent)	Tokyo							4,172			1 Feb. 1898	Transport difficulties in bad weather
2	Hokkaidō Timber Co.	Hokkaidō	8 Mar. 1899		Class 2			6,000				24 Aug. 1899	A fire
			12 Jan. 1900	30 Sep. 1900	Class 2			23,000	7,182			13 Dec. 1900	A shortage of ships
3	Sōtarō Shirato (Satarō Tsushima's agent)	Aomori	19 Oct. 1895					10,000	0			10 Dec. 1895	A breach of contract by a subcontractor a problem of raising money

(continued)

Table 2.1 (continued)

No.	Sleeper dealer	Address	Date of contact	Deadline for completion delivery	Tree varieties	Price of sleepers contracted (in yen)	Number of sleepers contracted	Number of sleepers delivered	Sleeper producing regions	Sleeper delivery destination	Petitions for cancellation of delivery contracts: Date of submission	Reason
4	Kanjūrō Nishimura	Aomori	3 Mar. 1896	Apr. 1896 May 1896			10,000	2,000			19 May 1896* 28 Apr. 1896	A shortage of freight cars due to military transports
5	Yasusuke Kikuta	Fukushima	9 June 1897	30 Sep. 1897	Class 1 (for switches)		1,280		Fukushima	Fukushima Station	28 Dec. 1898	A flood and transport difficulties because road was destroyed.
6	Toyokichi Numa-hata	Aomori	7 July 1899	30 Sep. 1899	Class 1	1,947	5,500		Iwate	Aomori	17 Dec. 1899	A shortage of freight cars, and a flood
7	Sakuzaem on Echigo	Akita	14 Oct. 1899		Class 1 (for switches)		2,682	973			22 Sep. 1900	Personal reasons
8	Takesuke Kakizaki	Akira	26 Jan. 1900		Class 1 (for switches)		1,713				28 Dec. 1900	Miscalculation of a unit sleeper price at competitive tendering process, and transport difficulties due to heavy snow

No.	Name	Address										Reason
9	Unokichi Uga	Fukushima	24 Oct. 1900	30 Dec. 1900	Class 1		5,000		Akita		27 Apr. 1900	Sinking of ship, and shortage of timber for sleepers
10	Miyota Horochi	Iwate	23 May 1899	30 Sep. 1899	Class 1 (for bridges)	3,900	2,000	572			29 Mar. 1901	A shortage of timber for sleepers
11	Tahichi Nozawa	Kanagawa	10 Oct. 1894			1,750	5,000	589	Iwate	Fukushima	22 Feb. 1900	An epidemic of dysentery
12	Tsunejirō Misawa (Kikujirō Misawa's agent)	Saitama	June 1894	Dec. 1894 → Sep. 1895		778	10,000	6,695		Tsuruga	4 June 1895	
							10,000				20 Sep. 1895	A shortage of ships due to military transports
13	Kirizaburō Yokoi (Kyōji Hasegawa's agent)	Tokyo	1 June 1900		Class 1	5,800	10,000	1,165			18 Dec. 1900	Personal reasons
14	Kidokoro	Tokyo	7 Nov. 1900		Class 1		5,000				23 Mar. 1901	Difficulties in carrying logs
15	Kakuzō Ishiwatari	Tokyo	Oct. 1896	30 Nov. 1896			402				29 Nov. 1896	A cancellation of a contract by a supplier; sleepers of crude quality

(continued)

Table 2.1 (continued)

No.	Sleeper dealer	Address	Date of contact	Deadline for completion delivery	Tree varieties	Price of sleepers contracted (in yen)	Number of sleepers contracted	Number of sleepers delivered	Sleeper producing regions	Sleeper delivery destination	Petitions for cancellation of delivery contracts Date of submission	Reason
16	Kamekichi Ichikawa	Tokyo	26 Nov. 1896		Class 1		5,000				9 May 1897	Natural disasters
			26 Nov. 1896		Class 2		5,000				31 Mar. 1897	Natural disasters
17	Yoshimasa Suzuki	Kanagawa	16 Dec. 1896	31 Mar. 1897	Class 2		7,700					
	Kamakichi Ichikawa (Yoshimasa Suzuki's agent)	Tokyo									25 Mar. 1897	Natural disasters
18	Yoshirō Yamaguchi	Tochigi	7 Feb. 1899			2,100	5,000				14 May 1900	
			22 May 1899	Aug. 1899	Class 1	2,250	5,000	4,689			6 Aug. 1900	An epidemic disease and a flood disaster
19	Tadajirō Nakamura	Yamanashi	20, 22 Jan. 1897		Class 2		20,000	3,378			31 Aug. 1897	Transport difficulties by road destroyed due to heavy rain

20	Kōshirō Nagasawa	Yamanashi	15 Oct. 1900	31 Dec. 1900	Class 1		15,000		Ibaraki		18 Mar. 1901	A reduction of water volume in the river
			15 Nov. 1900				20,000				15 May 1901	A reduction of water volume in the river
21	Takajirō Arai	Gunma		31 Jan. 1900	Class 1 (for switches)		686				9 May 1900	Heavy snow
22	Susumu Kōda	Nagano	15 June 1896	Nov. 1896		5,400	10,000	7,332		Shinbashi and Karuizawa Stations	26 Nov. 1896	A shortage of freight cars
23	Tsunezō Shimada	Nagano	12 Dec. 1899	31 Mar. 1900	Class 1		5,000				21 May 1900	Logging and transport difficulties due to heavy snow
24	Kinosaku Fujiwara	Nagano	21 Dec. 1900	10 Dec. 1900	Class 1 (for switches)						14 Mar. 1901	Personal reasons
25	Chōjirō Nishiwaki	Gifu	1 June 1900	31 Aug. 1900			20,000				1 Sep. 1900	Supplier's disease
26	Kumema Ōkura	Tokyo	20 Mar. 1901								5 July 1901	(Partial cancellation)
	Shōsaku Kanamori (Kumema Ōkura's agent)	Ngano		May 1901	Class 1 (for switches)						10 June 1901	Sleepers of crude quality

(continued)

Table 2.1 (continued)

No.	Sleeper dealer	Address	Date of contact	Deadline for completion delivery	Tree varieties	Price of sleepers contracted (in yen)	Number of sleepers contracted	Number of sleepers delivered	Sleeper producing regions	Sleeper delivery destination	Petitions for cancellation of delivery contracts Date of submission	Reason
27	Takeshirō Kameyama	Aichi	9 Aug. 1893	May 1894			5,000	More than half			20 June 1894	Personal reasons
28	Tōbē Okada	Aichi	25 Jan. 1901				20,000						
	Isaburō Nishiura (Tōbē Okada's agent)										8 June 1902	A shortage of timber for sleepers
29	Kintarō Umino (Bunkichi Idei's agent)	Shizuoka	3 Feb. 1897	24 June 1897			15,000				30 June 1897	Difficulties in logging
30	Kenkichi Ueno	Toyama	12 Nov. 1896		Class 1		5,000	1,500			30 Apr. 1897	The subcontractor didn't transport sleepers under the pretense of wind and waves in the sea
31	Kanejirō Kobayashi	Niigata	May 1896				1,300			Karuizawa		

	Shōsaku Tanaka (Kanejirō Kobayashi's agent)	Niigata	28 Mar. 1899	30 Sep. 1899	Class 1	5,000			21 Sep. 1899	Personal reasons
32	Sadakichi Kasai	Toyama	12 Feb. 1897			20,000	Akita	Fukui		
	Shinji Shirasawa (Sadakichi Kasai's agent)								30 Apr. 1897	Breach of contract by the sub-contractor
33	Buhē Ishida	Ishikawa	30 July 1897	30 Oct. 1897		5,000	Shiga		30 Oct. 1897	Difficulties in sawing due to a run of wet weather, and epidemic of dysentery
34	Hidekazu Katsumi	Ishikawa	6 Nov. 1899	25 Dec. 1899 → 29 Jan. 1900	Class 1 (for bridges)	1,141	Toyama, Ishikawa, Aichi, Fukui	Takaoka, Kanazawa, Tsuruga, and Kata-machi Stations		Logging and transport difficulties due to heavy snow
	Ichijirō Oda (Hidekazu Katsumi's agent)	Iishikawa							18 Feb. 1900*	

(continued)

Table 2.1 (continued)

No.	Sleeper dealer	Address	Date of contact	Deadline for completion delivery	Tree varieties	Price of sleepers contracted (in yen)	Number of sleepers contracted	Number of sleepers delivered	Sleeper producing regions	Sleeper delivery destination	Petitions for cancellation of delivery contracts	
											Date of submission	Reason
34'	Hidekazu Katsumi	Ishikawa	6 Nov. 1899	25 Dec. 1899	Class 1 (for bridges)	2,603	1,162					Logging and transport difficulties due to heavy snow, and wrong size of sleepers
	Sōemon Osa (Hidekazu Katsumi's agent)	Ishikawa									15 Feb. 1900	
35	Masayasu Aoki	Ishikawa	7 June 1899	26 July 1899	Class 1 (for bridges)		504				13 Aug. 1899	A fire
36	Hidejirō Tsuda (Yoshiharu Nagae's agent)	Ōsaka	23 July 1900			6,400	10,000	3,907			14 Mar. 1901	Transport and logging difficulties in bad weather and heavy snow
	Hidejirō Tsuda (Sentarō Kozasa's agent)		25 Aug. 1900				20,000				14 Mar. 1901	

No.	Name	Prefecture	Date	Deadline	Class	Qty	Qty	Location	Petition date	Reason
37	Kichibē Takagi	Hyōgo	6 Nov. 1900	31 Mar. 1901	Class 1	2,295	3,000		20 May 1901	Logging and transport difficulties due to heavy snow
38	Kazuyoshi Okada	Hiroshima	13 Feb. 1899		Class (only chestnut)		5,000	Ōsaka	31 Mar. 1899	Sleepers of crude quality
			20 Feb. 1899		Class (only chestnut)	2,285	5,000		30 Apr. 1899	Sleepers of crude quality
39	Kichitarō Abe	Tottori	4 Mar. 1901		Class 1		6,984		1 Mar. 1902	Natural disasters
					Class 1	7,450	10,000			
40	Kōzō Minami	Tottori	5 Oct. 1900				10,000	Shimane	9 Mar. 1901	Mistakes in conclusion of contract

Source: Teishinshō (1893–1902)

Note: (1) * a petition for postponement of the deadline for delivery completion. (2) Class 1 timber: chestnut, hiba and hinoki. Class 2 timber: timber from Hokkaidō

and the construction of the railways between Hachiōji and Nagoya and between Shinonoi and Shiojiri, which began in that year. Tendering contests were held at accounting offices in Shizuoka, Kōbe, Nagano, and other regions that were under the jurisdiction of the Accounting Division of the Ministry of Communications' *Tetsudōkyoku* (Dajōkan Monjokyoku 1903). To complete the delivery process, *Tetsudōkyoku* had the winning dealer bring sleepers to a specified location at the site of a railway station by a given deadline and then conducted a product quality inspection. Although there are some exceptions, the number of sleepers contracted was 5,000–20,000 (5,000 per lot) per dealer. A calculation of the number of dealers who delivered sleepers based on the number of sleepers purchased by *Tetsudōsagyōkyoku* in 1899 and 1900 reveals that there were at least 22, and at most 89, such dealers in 1899 and at least 39, and at most 157, dealers in 1900. Since the number of the dealers who petitioned for contract cancellation is 10 (14 instances) for 1899 and 12 (15 instances) for 1900 as shown in Table 2.1, it is estimated that *Tetsudōsagyōkyoku* signed contracts with 60–100 sleeper dealers, and that about 10 % of them submitted petitions for contract cancellation. The proportion of the amount paid for the sleepers procured though competitive tendering to the total amount spent for all sleepers is 97 % (216,411 yen) for 1899 and 99 % (586,433 yen) for 1900 (Tetsudōsagyōkyoku 1900).

3.3 Limitations Placed on the Participants in the Tendering Process

Under the competitive tendering system, the contract winners were determined based only on the tendering price, and the penalty for non-execution of the contract was only the confiscation of the security deposit, which gave rise to problems such as the participation of financially and technologically weak producers in tendering contests and the delivery of poor quality sleepers. Although the reasons for contract cancellation listed in Table 2.1 include natural disasters, the number of sleepers that had been planned for delivery in 1899 but were not actually delivered was 67,000, or about 15 % of the number of sleepers purchased by the Ministry of Communications' *Tetsudōsagyōkyoku* that year (447,000). In response, with the issuance of Ordinance of the Ministry of Communications No. 19, the Ministry put limitations on the participants in the tendering process based on the amount of their direct national tax. More specifically, participants with a tendering price of less than 5,000 yen per lot (5,000 or more but less than 10,000; 10,000 or more but less than 20,000; 20,000 or more but less than 50,000; and 50,000 or more) must be taxpayers paying a national direct tax of 10 yen (20, 50, 80, and 100, respectively) or more annually. Also, one employee in the case of a general partnership company or one partner with unlimited liability in the case of a limited partnership company must satisfy the above condition. In the case of a stock company, it must have the same qualification as a limited partnership company, or the money received for issued stocks must be twice as much or more than the quoted price (Naikaku Kanpōkyoku 1900). With such limitations, the Ministry of Communications determined the quantity of sleepers delivered based on the suppliers' asset size and lowered the risk of contract non-execution in efforts to secure sleepers accordingly to the specifications, tree varieties, and quantities listed on the contracts.

An invited tendering system was created in June 1900 with the issuance of Imperial Ordinance No. 280. This enabled the Ministry of Communications' *Tetsudōsagyōkyoku* to directly select the participants for the tendering process. As a result, things like coal, cement, girders, and materials for train vehicles were purchased through the invited tendering system (Teishinshō 1904, 1905). The competitive tendering system continued to be used for sleepers however. One reason may be that producers were unlikely to meet delivery deadlines even if they had started producing sleepers immediately after signing their contract; production of sleepers required more than half a year for logging, transportation, sawing, and drying. However, the delivery deadline was set to be 3–5 months after the contract was signed. Further, despite the fact that logging and transportation in some regions was difficult due to seasonal reasons such as accumulated snow, sleeper dealers could not know the quantity to be contracted or the delivery location and delivery deadline until immediately before the tendering contest, which made planned production impossible. Another reason why competitive tendering for sleepers continued is that since the transportation costs were high for a distant delivery location, the geographical extent of sleeper-producing regions for a given delivery location was, to some degree, limited. For these reasons, the Ministry of Communications' *Tetsudōsagyōkyoku* would likely have experienced difficulties securing the necessary quantities without opening up the tendering system to many sleeper dealers, although eliminating those who paid less than 10 yen for their direct national tax.

In sum, with the competitive tendering process or the invited tendering process, it was difficult to secure the necessary quantities of sleepers while keeping their quality high.

4 Post-nationalization Period Until the Start of World War I

4.1 Procurement of Sleepers Through Sole-Source Contracts

New regulations wereenacted in June 1906 (and implemented the following year) on the Imperial Railway and the accounting of funds for materials; accordingly, the procurement process for railway materials shifted from competitive tendering to sole-source contracts. As seen in Table 2.2, the share of sole-source contracts in the amount spent for purchasing materials (domestic orders only) increased from 22 % in 1906 to 58 % in 1907 and then to 74 % in 1909. The Railway Authority (Tetsudōin 1908–1920) came to use sole-source contracts, starting with sleepers purchased for 1909 (Teikoku Tetsudō Taikan Hensankyoku 1984). It also took the following measures to secure the necessary quantities of sleepers.

Table 2.2 The amount spent by JGR in purchasing materials, 1900–1909

Year	Domestic orders					Foreign orders	Total (in 1,000 yen)
	Competitive tendering (%)	Invited tendering (%)	Sole-source contract (%)	Others (%)	Sub-total (%)		
1900	7,362 (88.9)	99 (1.2)	822 (9.9)	0 (0.0)	8,283 (100)	3,798	12,081
1901	3,636 (59.6)	570 (9.3)	549 (9.0)	1,344 (22.0)	6,099 (100)	1,767	7,866
1902	3,402 (59.6)	1,695 (29.7)	611 (10.7)	0 (0.0)	5,708 (100)	3,019	8,727
1903	2,753 (35.0)	3,682 (46.8)	1,429 (18.2)	8 (0.1)	7,872 (100)	1,834	9,706
1904	1,073 (41.3)	739 (28.4)	707 (27.2)	82 (3.2)	2,601 (100)	2,263	4,864
1905	1,668 (24.4)	3,749 (54.7)	1,420 (20.7)	11 (0.2)	6,848 (100)	645	7,493
1906	4,363 (41.5)	3,863 (36.7)	2,270 (21.6)	29 (0.3)	10,524 (100)	2,124	12,648
1907	9,558 (34.8)	1,849 (6.7)	15,980 (58.2)	63 (0.2)	27,449 (100)	7,931	35,380
1908	5,012 (25.3)	3,846 (19.4)	10,898 (55.1)	31 (0.2)	19,788 (100)	2,073	21,860
1909	2,891 (13.2)	1,116 (5.1)	16,316 (74.4)	1,622 (7.4)	21,944 (100)	2,120	24,064

Sources: Tetudōsagyōkyoku (1900–1905); Tetsudōchō (1906 and 1907); Tetsudōin (1908 and 1909)

Note: Others include contracts with the Yawata Steel Works and the Printing Bureau

The Railway Authority decided to limit the scale of material purchases in different regions (at railway management bureaus and regional offices) and to make lump-sum purchases in Tokyo (at the headquarters). It totally banned regional purchase of coal and sleepers, for which the amount spent was particularly large. In addition, the Railway Authority conducted investigations related to purchasing special materials, such as sleepers and coal, and to preparing purchase plans. Although details of the investigations related to sleeper purchases are unknown, after the contracts in a given year had been completed, the Railway Authority sent staff to sleeper-producing regions to collect information on unit prices and production quantities for the following year and conducted detailed investigations and research. Then, based on these investigations, the Authority judged whether the production costs (calculated precisely based on the price of timber, production costs, transportation costs, and profit) listed in the quote form were high or low, calculated the quantity to be contracted for—which would ensure the securing of the necessary quantities within budget—and gave instructions to the designated sleeper dealers. The quantity desired by a dealer was not listed on the quote form. By informing dealers of its desired contract quantity after receipt of the quote forms, the Railway Authority probably intended to imply that the contract quantity would be reduced if a sleeper dealer's quote was relatively high, and in this way would prevent them from raising the prices quoted. After informing the sleeper dealers of its desired contract quantity, the Railway Authority decided on the contract unit price through price negotiations and prepared a final purchase plan. The contract price determined in this way was not changed over a one-year contract period unless there was a significant change in the market price (see Fig. 2.3) (Nihon Kokuyū Tetsudō 1971; Nihon Makuragi Kyōkai 1959).

The Railway Authority selected sleeper dealers who owned mountains and forests and had sufficient funds and credit as transaction counterparts, and decided

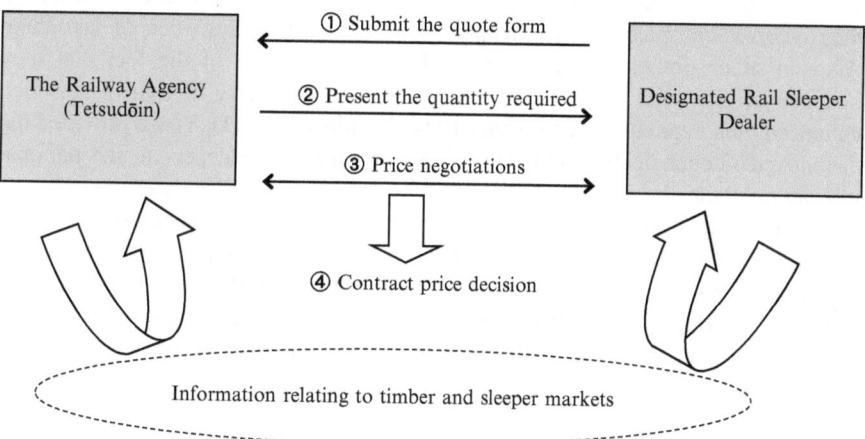

Fig. 2.3 The process of concluding contracts for the purchase of sleepers

to make direct contracts with them (Yamada 1911). The criteria used to select designated sleeper dealers are unknown, but, for example, Sōbei Suzuki of Zaisō and Tadashichi Hasegawa of Hasegawa Shōten served as executives at many timber companies and chambers of commerce and were powerful timber dealers in Nagoya with sizable assets. Sannosuke Kobayashi, whose assets and volume of timber transactions were probably small relative to Suzuki's and Hasegawa's, opened a sleeper dealing business in Ichikawachō, Kanzakigun, Hyōgo prefecture in 1908, then moved to Gifu prefecture in 1916 and started producing and selling sleepers mainly in that prefecture, and eventually became one of the Ministry of Railways' (Tetsudōshō 1920–1943) designated sleeper dealers in 1921 (*Gifu Shinbun*, 3 December 1995).

The Railway Authority not only made contracts with such designated sleeper dealers and did not easily allow new entry by other dealers, but also rarely made contracts with sleeper dealers who had previously caused a contract cancellation due to contract non-execution. Yoshimoto Shōten (established in 1887; Yoshimoto Limited Partnership Co. established in March 1910 with a capital of 5,000 yen) was a timber dealer headquartered in Saku, Nagano. In addition to producing wood charcoal, it also produced sleepers as side work and started to deliver sleepers to the Nagano Transportation Office of the Ministry of Communications' *Tetsudōsagyōkyoku* in 1905. Starting in the mid-1910s, Yoshimoto Limited Partnership Co. expanded its sleeper sales to several private railway operators including Nagano railways, Ina railways, and Tōbu railways, and the number of sleepers delivered well exceeded 100,000 per year. However, the delivery of sleepers from the company to the Ministry of Railways stopped in 1921 for an unknown reason. The company subsequently petitioned the Ministry of Railways repeatedly for the right to supply sleepers, but permission was not granted (Yui 1961). Such actions by the Railway Authority or the Ministry of Railways guaranteed that, in exchange for proper contract execution, designated sleeper dealers would receive contracts the following year. Therefore, the designated sleeper dealers could secure stable sales. In addition, they received the payment for sleepers delivered within one week of invoicing. Also, in other timber markets they could take advantage of the fact that they had transactions with the Railway Authority or the Ministry of Railways, which enhanced their reputation (Tomiyama 1934; Yoshitsugu 1951). These provided the designated sleeper dealers with incentives for supplying sleepers to the national railway operator.

The Railway Authority had sleeper dealers including Kyōji Hasegawa (Tokyo), Torazō Fuji (Ōsaka), Kinzaburō Nagata (Nagoya), and Yasujirō Kodate (Aomori) establish the Ōminato Timber Co. in Aomori in August 1912 in order to secure sleepers for bridges and switches, for which mostly only hiba and hinoki timber could be used in Honshū.[4] With a starting capital of 500,000 yen (125,000 yen

---

[4]According to a survey of the income and business taxes of September and October 1913, the number of timber dealers involved in the establishment of Ōminato Timber Co. was relatively large (Kōjunsha ed. 1914).

already paid in), the company set up its headquarters in Ōminato, Aomori and an office in Fukagawa, Tokyo and started to produce sleepers, with preferential purchases of hiba timber from national forests in Aomori. In the 1900s, government-run timber factories in Aomori and Nagano produced sleepers, using timber from national forests, but there was an increasing movement toward their abolishment in the first half of the 1910s because they took business away from private companies or did not perform well. Under these circumstances, the Railway Authority gave up on the management of sleeper factories, which was under negotiation with the Ministry of Agriculture and Commerce (Nōshōmushō 1910), and instead commissioned private-sector actors to establish sleeper-producing companies (Matsunami 1924; Hasegawa Mokuzai Kōgyō Kabushiki Kaisha 1967). It seems that the Nagoya Railway Sleeper Limited Partnership Co., which was established in May 1913 with a capital of 120,000 yen, served the same function as the Ōminato Timber Co. and produced sleepers for bridges and switches, purchasing hinoki timber from the Kiso Imperial Forest (*Goryōrin*). The investors in the Nagoya Railway Sleeper Limited Partnership Co. were Sōbei Suzuki, Kinzaburō Nagata, Tadashichi Hasegawa, Kōjuro Hattori, Kentarō Hayase, and the Hamakiya Limited Partnership Co., all powerful timber dealers in Nagoya (Nagoya Shōgyō Kaigisho 1914; Nihon Kōtsū Kyōkai 1952).

Thus, the Railway Authority selected sleeper dealers with sufficient funds and credit and made contracts with them at prices that were judged to ensure the securing of sleepers. With this system the Railway Authority could secure the necessary quantities of sleepers within budget, avoid regional shortages of sleepers caused by differences between the existing quantity of timber and demand for it, and ensure the quality of sleepers.

4.2 Activities of the Designated Sleeper Dealers

What kinds of activities were conducted by the designated sleeper dealers in connection with the procurement of sleepers by the Railway Authority? This subsection examines the activities of Hasegawa Shōten which supplied sleepers to JGR. The Hasegawa family lived in Shimo-Asōmura, Kamogun, Gifu prefecture from the 1680s. The head of the family served as the village headman, and the family conducted a log boom project and a logging project in the Meiji period. The Hasegawa family collected logs floating downstream from Hida and Gujō with a log boom at Shimo-Asō, formed them into rafts, and transported them to Nagoya, Kuwana, and Inuyama. Since Nagoya, a major destination for timber, was an important market, the family set up a branch store in Kuwana in 1876 as a distribution centre for timber and another branch store in Nagoya in 1881. The family also set up its Tokyo office in 1881 (upgraded to branch status in 1886) and its Ōsaka branch in 1892, expanding its sales network to both these cities. Among the three branches in Nagoya, Tokyo, and Ōsaka, the Nagoya branch supplied timber not only to the Navy and the Railway Authority, but also to companies engaged in civil engineering construction, spinning, and harbour construction. The Nagoya branch apparently performed better than the Tokyo branch and the Ōsaka branch (Hasegawa Mokuzai Kōgyō Kabushiki Kaisha 1967; Hasemoku 1988).[5] According to a survey conducted in September and October of 1910, the business tax paid by Tadashichi Hasegawa (head of the Nagoya branch), Kyōji Hasegawa (head of the Tokyo branch), and Katsusuke Hasegawa (head of the Ōsaka branch), was 892 yen, 312 yen, and 339 yen, respectively (Kōjunsha ed. 1911).

Since the inception of the Nagoya branch, its head, Tadashichi Hasegawa, actively purchased mountains and forests in nine prefectures including Gifu, Nagano, Aichi, Shizuoka and Mie, and set up offices for logging and transportation. In light of this, the Hasegawa family created a central command center within the Nagoya branch in 1895. The central command center managed the family assets and business operation, supervised the operation of the branches, and unified administrative work related to logging projects and shipping. Through the headquarters and the Nagoya branch, the command centre conducted logging and transportation projects, timber sales, and investments in forests and mountains. However, sales were stagnant relative to timber production before World War I. In October 1913, the command centre was shut down and all branches began to operate autonomously (Hasemoku 1988).

Let us now examine in detail the business activities of the Hasegawa Tokyo branch. Against a backdrop of increased demand in Tokyo for building materials and materials for civil engineering construction to develop transportation infrastructure, the Tokyo branch engaged in the consignment sale of timber and, at the same time, focused on direct sales of timber produced by its family business. However, for several years after its establishment in 1887, the Tokyo branch could not join the associations organized by the wholesale sellers and buyers of timber, and its business was so bleak that it was about to face bankruptcy. In response, the Hasegawa headquarters decided to separate the accounting system of the Tokyo branch from that of the headquarters and made the branch operate independently after providing it with a capital fund of 10,000 yen in 1892. It was then that the Sino-Japanese War broke out, which led to an economic boom and increased demand for building materials. This enabled the Tokyo branch to turn around its declining business. As seen in Table 2.3, which shows the operational performance of the Tokyo branch, it earned profits of 1,075 yen and 3,923 yen in 1894 and 1895, respectively. However, during the post-Sino-Japanese War economic stagnation, demand for timber fell. Also, since the headquarters had increased its loan to the Tokyo branch by 10,000 yen after the account was settled for 1897, the interest on the capital from the headquarters increased. These factors led to the Tokyo branch posting a loss in 1898. Therefore, in order to stabilize its business, the branch expanded its timber supply to government agencies, taking advantage of the fact that it had supplied timber to the Imperial Household Ministry (*Kunaishō*) (Hasegawa

---

[5]The Kuwana branch was closed in 1890.

Table 2.3 Profit and loss statement of the Hasegawa Tokyo branch, 1894–1916

Year	Revenues					Costs and expenses						Gross profit (C = A−B)	Interest on capital from headquarters (D)	Net income (C−D) (in yen)
	Sales of goods	Sales commission fee	Interest income	Other income	Total (A)	Cost of goods sold	Selling, general and administrative expenses	Interest expenses	Rent expenses/taxes expenses	Other expenses	Total (B)			
1894														1,075
1895	36,638	4,247	501	496	41,882	31,818	3,973	278		1,224	37,292	4,590	667	3,923
1896	65,019	6,867	2,165	308	74,359	56,106	5,345	665		1,674	63,790	10,569	600	9,969
1897	47,326	1,017	1,497	7,677	57,517	40,674	5,577	260		1,647	48,158	9,359	1,204	8,156
1898	38,835	491	1,336	1,086	41,748	36,057	3,728	453		158	40,395	1,352	1,460	▲108
1899		1,992	1,372	2,389			7,211	1,002		2,121		5,095	1,464	3,631
1900	94,692	2,450	429	2,104	99,674	76,569	10,780	2,805		520	90,676	8,998	1,670	7,328
1901	108,863	1,860	1,069	935	112,726	87,892	14,319	3,463	814	0	106,487	6,239	480	5,750
1902	79,739	2,652	952	5,023	88,366	71,820	4,417	2,499	851	0	79,586	8,780	2,499	6,281
1903	106,673	4,057		3,872	114,602	98,020	5,116	3,277	1,123	422	107,959	6,644		6,644
1904	134,945	1,017		4,909	140,871	120,761	8,676	813	1,585	0	131,835	9,036		9,036
1905	245,236	6,518		3,350	255,104	222,509	13,709	429	1,922	748	239,316	15,788		15,788
1906	244,494	7,373		12,688	264,555	228,573	19,419	1,823	2,684	5	252,504	12,051		12,051
1907	210,503	861		3,433	214,797	188,282	17,482	0	2,864	0	208,629	6,168		6,168
1908	261,046	3,304		1,288	265,639	237,166	16,356	4,713	1,396	0	259,630	6,008		6,008
1909	168,497	1,385		0	169,882	144,747	16,755	1,947	1,518	1,931	166,899	2,984		2,984
1910	131,656	160		4,180	135,996	112,952	12,881	3,298	286	1	129,418	6,578		6,578

(continued)

Table 2.3 (continued)

Year	Revenues					Costs and expenses						Gross profit (C = A-B)	Interest on capital from headquarters (D)	Net income (C-D) (in yen)
	Sales of goods	Sales commission fee	Interest income	Other income	Total (A)	Cost of goods sold	Selling, general and administrative expenses	Interest expenses	Rent expenses /taxes expenses	Other expenses	Total (B)			
1911														
1912	176,649	0		1,826	178,475	147,025	17,061	4,477	940	1,474	170,977	7,498		7,498
1913	178,674	2,989		3,520	185,182	159,772	14,450	3,488	851	0	178,561	6,621		6,621
1914	191,262	5,354		5,647	202,263	175,124	15,019	2,237	794	4,518	197,692	4,571		4,571
1915	109,352	1,798		2,594	113,745	92,139	15,618	1,805	799	0	110,361	3,384		3,384
1916	140,806	765		2,917	144,488	119,059	13,179	4,467	779	0	137,483	7,005		7,005

Source: Hasegawa Tokyo Shiten (1895–1910, 1914 and 1915)
Notes: The accounting period is from January to December. ▲ means minus

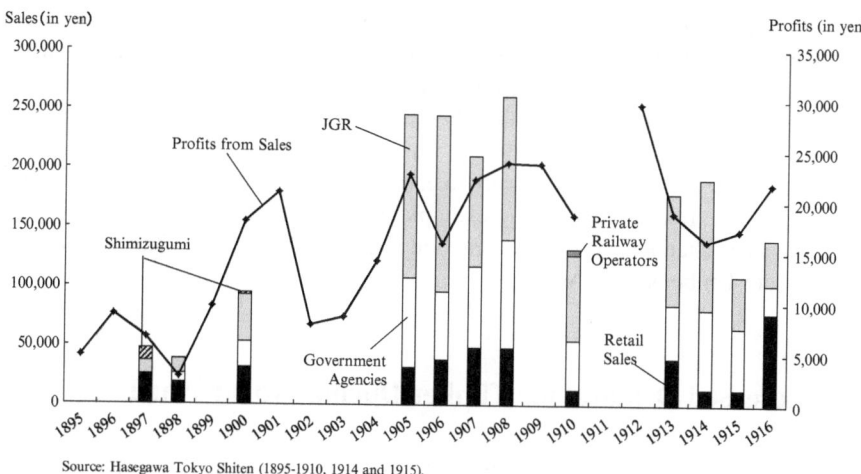

Source: Hasegawa Tokyo Shiten (1895-1910, 1914 and 1915).

Fig. 2.4 Sales and profits of products of the Hasegawa Tokyo branch, 1895–1916

Mokuzai Kōgyō Kabushiki Kaisha 1967). The profit of the Tokyo branch mainly consisted of the profit from product sales (subtracting the cost of goods sold from the sales of goods) and sales commission fees. From 1900, of the entire profits, the share of profit from product sales was largest, at 60–90 %.

Figure 2.4 shows product sales of the Tokyo branch from 1895 to 1916. The branch engaged in timber supply to government agencies (JGR, Navy Shipyard, The Ministry of the Imperial Household, the Water Supply Department of Tokyo City, and the agency overseeing civil engineering works), timber sales to Shimizugumi (Shimizu Construction Co.) and private railway operators, and retail sales. Sales to government agencies were the largest; these accounted for 25 % of all sales in 1897 and 51 % of all sales in 1898. These sales reached 80–90 % of all sales from 1905 to 1915. In particular, the share of sales to JGR in the entire product sales was around 50 % from 1905 to 1914, except for the years for which data are not available, and exceeded 60 % in 1906. Besides timber supply to government agencies, retail sales were important. The share of retail sales in the product sales varied significantly from 7 to 50 %, but the fluctuation coincided with that of the profit from product sales.[6] For the Hasegawa Tokyo branch, timber supply to JGR was likely to cause a loss when there was a post-contract increase in the price of trees, logs, and labour, but was important during a recessionary period when demand for general-purpose timber declined. At the same time, in retail sales, it

---

[6] As an exception, in 1905, although the retail sales declined from the previous year, the net profit rose by more than 3,500 yen. This can be attributed to the fact that during the Russo-Japanese War, when the timber price was increasing due to rapidly increased demand for materials for military purposes, the Tokyo branch sold timber, which it had purchased before the war and had stored, to the Navy and Army.

was difficult to achieve distribution network expansion or sales increases during economic stagnation, but it was feasible during periods of economic boom when timber demand rose. In other words, for the Tokyo branch, the sleeper market was a sector where the branch could definitely secure a certain level of profits each year, and the retail market was a sector which was affected by economic fluctuations, but which gave the branch an opportunity to earn large profits during economic booms. The Hasegawa Tokyo branch is an example of a designated sleeper dealer having multiple sales channels. The same situation was also observed for Zaisō in Nagoya and Kodate Timber Co. in Aomori.

5 Increased Demand for Timber During World War I

As discussed in the previous section, the Railway Authority procured sleepers through designated sleeper dealers. However, during the economic boom resulting from World War I, it became difficult to secure stable supplies of sleepers. After the Japanese economy recovered from the recession in the second half of 1914, a wartime boom resulted from increased exports to Asia and the United States. Industrial activity was invigorated in various sectors, increasing the demand for timber. As a result, the price of general-purpose timber soared, and the average unit price for regular sleepers paid by the Railway Authority rose from 0.62 yen in 1915 to 1.04 yen in 1917, then to 1.79 yen in 1919. In early 1917, the Railway Authority could sign contracts at almost the same price as the previous year, with traditional direct procurement from the designated sleeper dealers; however, as wages increased from the middle of the year, many dealers asked for delayed delivery or contract cancellation, which hindered the progress of projects. Therefore, as an incentive payment, 20 % of the contract price was granted for half of the contract quantity (i.e., 0.1–0.2 yen per sleeper) so that the Railway Authority could then procure the necessary quantity (Tetsudōin 1916–1919). Such a problem can be attributed to the fact that the designated sleeper dealers were not specialized in the production and sale of sleepers for the Railway Authority, and that as demand for timber increased, they sold timber in other timber product markets, where they expected to earn greater profits, thereby reducing the quantity of sleepers supplied to the Railway Authority or cancelling contracts.

For example, in Hokkaidō, where there was active exporting of sleepers, the price of sleepers for exports (about 8 ft) was high compared to the price of sleepers supplied for railways in Hokkaidō or Honshū. Therefore, timber dealers preferred to produce and supply 8-ft sleepers intended for railways in Korea and Manchuria, resulting in chronic shortages of sleepers for local railways. Particularly in 1916 and 1917, it was enormously difficult for the railway operator in Hokkaidō to procure sleepers. In response, the Railway Authority significantly increased the budget for sleeper procurement and forcibly secured delivery of the necessary quantity for 1918 (approximately 500,000 sleepers) to the railway management bureau in Sapporo

(*Kōbe Yūshin Nippō*, 16 October 1918). As a result, the purchase price per sleeper paid by the railway management bureau increased—from 0.40 to 0.45 yen in 1916, from 0.45 to 0.55 yen in 1917, and from 0.65 to 0.70 yen in 1918. However, as the price of a sleeper for export (at Otaru port) rose from 1.25 yen in 1918 to 1.90 yen in 1919 and to 2.25 yen in 1920, the price difference between sleepers supplied to the railway management bureau in Sapporo and sleepers for exports grew to about threefold, and there were many dealers who had signed a contract and had paid their security deposit to the Railway Authority but later cancelled the contract and paid the specified penalty (*Hokkai Times*, 28 December 1918). Especially for a timber dealer like Mitsui & Co. (Mitsui Bussan) that produced and sold timber on a large scale in both the domestic and overseas market, the supply of sleepers to the Railway Authority must have seemed small. Mitsui & Co. produced 2 million cubic meters of timber in Hokkaidō at the end of 1919, and 3.25 million of it was for sleepers intended mainly for exports to China and Korea. In Honshū, there was a case where 10,000 sleepers were produced from the mountains of Iwate prefecture and supplied to the Railway Authority in 1919; however, according to a report by the head of the Miike branch in a conference with Mitsui's branch chiefs, supplying sleepers to the Railway Authority was only a nominal business for Mitsui, yet it was advantageous because they received a certain amount of fees per tie without taking any risks (Mitsui Bunko 2004; Ringyō Hattatsushi Kenkyūkai 1958).

In the case of the Hasegawa Tokyo branch, 93,385 yen's worth of sleepers were supplied to the Railway Authority in 1913 and 110,075 yen in 1914, but the amount fell to 38,412 in 1916. In contrast, retail sales increased from 39,897 yen and 13,957 yen in 1913 and 1914, respectively, to 78,374 yen in 1916. Similarly, profits from product sales rose from 18,902 yen and 16,138 yen in 1913 and 1914 to 21,747 yen in 1916 (see Fig. 2.4). If a dealer canceled a contract with the Railway Authority, it was possible that transactions with the agency in the following years would not be guaranteed, and that sales channels could not be secured especially during a recessionary period when demand for timber declined. Therefore, it seems that the Hasegawa Tokyo branch, for which the share of sales to the Railway Authority in total sales was relatively large, reduced the quantity of sleepers supplied to the agency without cancelling contracts on the one hand, while at the same time taking advantage of its reputation as a supplier to the agency to pursue greater profits in other timber markets.

Facing difficulties securing sleepers because of the designated sleeper dealers' business savvy, the Railway Authority procured sleepers to be used in its Okayama railway district, for example, from Okayama, Tottori, Shimane, and Ehime prefectures, as well as relatively distant Nagano prefecture, in order to avoid regional sleeper shortages. The Authority brought in 200,000 sleepers in 1918, and 250,000 sleepers in 1919 from Hokkaidō to make up for shortages in Honshū. In contrast, the Tokyo railway management bureau could secure only 330,000 sleepers in 1920, as opposed to its annual demand for 600,000 sleepers (*Hokkai Times*, December 28, 1918; Dai-Nihon Sanrinkai 1920).

6 Shortages of Timber Suitable for Sleepers in the 1920s

Demand for timber fell in the 1920s due to the postwar economic recession. Although demand for timber rose rapidly during the period of World War I, after the war the government lowered tariffs on timber. The government also removed most of the tariffs on imported timber after the Great Kantō Earthquake in September 1923, seeing the need for timber in the reconstruction of the earthquake-stricken areas. As a result, the amount of imported timber jumped from 34,000 cubic meters in 1920 to 3.3 million cubic meters (of which 2.7 million cubic meters came from the United States) in 1924. At the same time, timber which was harvested through large-scale logging due to insect damage, was transported from Sakhalin. Japan's timber market therefore experienced excess supply, which led to a fall in the price of timber (Ōkurashō 1920–1925). During recessionary periods when demand for general-purpose timber declined, existing designated sleeper dealers supplied sleepers as contracted, so many timber dealers wanted to become designated sleeper dealers for the Ministry of Railways in order to secure sales channels. But in the 1920s the Ministry of Railways still had to tackle the problem of securing sleepers. This is because it was plagued by shortages of timber suitable for sleepers (appropriate tree varieties with a large diameter) due to the greater demand for sleepers resulting from increased railway construction and improvement work. Concerns about sleeper shortages had been expressed since railway nationalization, and chestnut timber was expected to be in short supply around the mid-1910s at the latest (*Yokohama Bōeki Shinpō*, 9 October 1909; Nōshōmushō 1910).

The sleepers used by the Ministry of Railways were divided into Class 1 timber (chestnut, hiba, hinoki, and Japanese chinquapin) and Class 2 timber (tree varieties used in Hokkaidō without preservation treatment, mainly timber from Hokkaidō). Facing shortages of suitable timber, the Ministry designated other tree varieties usable for sleepers as Class 3 timber. It also expanded the coverage of Class 3 timber, and the number of tree varieties designated increased from 9 and 17 in 1900 and 1910, respectively, to more than 40 by the second half of the 1930s (Suzuki 1938). However, while Class 1 timber lasted 8–12 years, Class 3 timber, such as pine and beech, lasted 3–5 years. As the replacement frequency was high for sleepers of Class 3 timber, the Ministry of Railways had to apply preservation treatment (inject creosote oil, etc.) to Class 3 timber. Some timber imported from the United States and Sakhalin was added to the list of the designated tree varieties, but tree varieties suitable for sleepers were very limited. Oregon pine, which was the typical variety imported from the United States, was used with creosote oil injected.

Research on the preservation techniques for timber started being actively conducted after railway nationalization in 1906, and the Operation Research Committee was set up after Shinpei Gotō took office as president of the Railway Authority in December 1908. The Committee's Eleventh Sub-committee was mainly responsible for research on sleepers and conducted comparative experiments on different preservatives, injection quantities, and injection methods. Also, in 1913, in order to conduct research on tree varieties suitable for preservation treatment, the Railway Authority set up a timber research laboratory, which was equipped with a room for testing preservatives, conducting analysis, and culturing bacteria, and began developing preservation techniques. The Railway Authority moved systems and equipment from the Sunagawa timber preservation treatment factory (established in 1909) of the Hokkaidō management bureau and the Utsunomiya preservation treatment factory (established by the Nippon Railway Co. in 1900 and transferred to the Railway Authority at the time of railway nationalization) to Fukagawa, Tokyo and started preservation treatment for sleepers at the new location in December 1919. The Railway Authority (the Ministry of Railways) applied preservation treatment to 310,000 regular sleepers and 2,400 sleepers used for bridges and switches in 1920, and produced 400,000 treated sleepers annually from 1926 through 1929. In the early 1910s, there were only two private-sector preservation treatment factories, which were owned by Nihon Wood Preserving Co. and Tōyō Wood Preserving Co., but the number of such factories increased to 12 by the mid-1920s. Partly due to this increase, the Railway Authority specified preservation treatment factories as the delivery site for sleepers supplied by dealers and had the sleepers processed there (Mokuzai Hozonshi Hensan Iinkai 1985; *Tetsudō Jihō*, 19 November 1921; Nihon Kōtsū Kyōkai 1952).

The unit price of treated sleepers produced and procured by the Ministry of Railways in this way was higher than that of Class 1 sleepers by 0.3–0.5 yen in the second half of the 1920s. Also, the useful life of treated sleepers was longer than that of Class 1 sleepers by only 2–3 years. Furthermore, due to a dearth of large-diameter trees and widespread use of small-diameter logs, the number of sleepers made of boxed timber, which tended to cause cracks and thus rotting, increased. This further decreased the cost performance of treated sleepers. The share of treated sleepers in all sleepers procured by the Ministry of Railways was around 30 % from the end of the 1920s through 1930. Therefore, the Ministry still needed the designated sleeper dealers to supply as many desirable sleepers as possible (Kamimura 1935; Tetsudō Jiho, 18 January 1930).[7]

7 Greater Shortages of Sleepers in the 1930s

7.1 Reduced Material Purchase Expenditure due to Austere Fiscal Policy

In addition to the problem of securing appropriate materials, the Ministry of Railways faced budget problems at the end of the 1920s. Each year, the Ministry prepared the following year's budget for three accounts—capital, revenue, and supplies—based on forecasts of the following year's operating income. The revenue items in the capital account were profits carried forward, bonds, and borrowings,

---

[7]It seems that the use of steel sleepers was difficult as the domestic supply of rails was finally achieved in the mid-1920s (Tomiyama 1934).

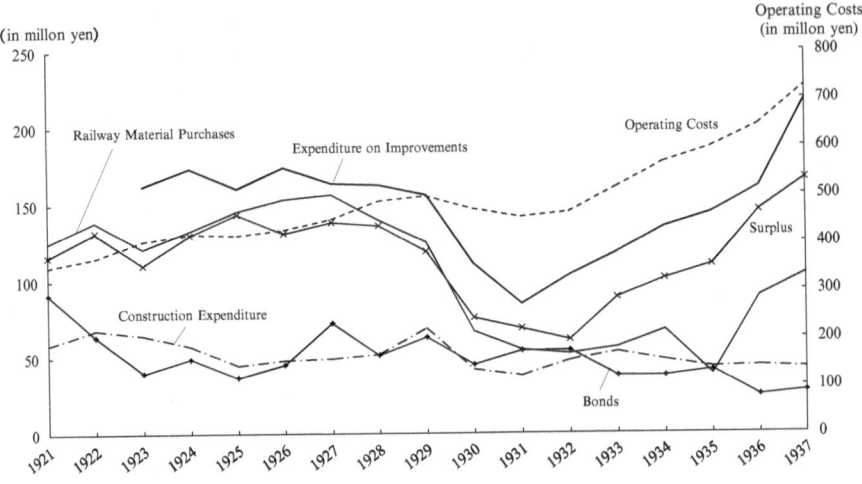

Sources: Tetsudōshō (1941), pp.22-30; Nihon Kokuyū Tetsudō (1971), pp.511-514.

Fig. 2.5 Trends in the operations of the Ministry of Railways, 1921–1937

and main expenditure items were construction expenditure, which was needed for construction work for new railway lines, and improvement expenditure, which was necessary for work for additional tracks or gradient improvement. The revenue items in the revenue account were transportation income and miscellaneous income, and the expenditure items were operating costs and subsidy costs. The difference between the revenue side and the expenditure side surplus was carried forward as revenue in the capital account and was mainly used for improvement expenditure. As shown in Fig. 2.5, among these items (all of which are of the category *kō*, one of the budgetary categories used under the then Public Accounting Act), construction expenditure declined from 58 million yen in 1921 to 45 million yen in 1924, but started to increase in the following year and reached 69 million yen in 1929. Improvement expenditure gradually increased from 125 million yen in 1921 to 156 million yen in 1928 and consistently exceeded construction expenditure in the 1920s. Operating costs increased from 349 million yen in 1921 to 497 million yen in 1929, and the total revenue in the revenue account was between 466 million yen and 623 million yen over the same period. Therefore, profit was between 111 million yen and 140 million yen. As a result, there was a funding shortage of 2.15 million to 54 million yen to cover improvement expenditure, which depended on surplus as the main funding source. Active railway construction and improvement became impossible under such funding deficiencies following the implementation of an austere fiscal policy in 1930. Expenditures for construction and improvement for 1930 significantly dropped to 41.71 million yen and 66.73 million yen respectively, and operating costs for the same year also fell by 27 million yen from the previous year (Nihon Kokuyū Tetsudō 1971).

Since the Ministry of Railways was not allowed to transfer funds between items of the category *kō*, it needed to respond to the budgetary shortfall for material purchases by either using stored products or reducing contract prices. The Ministry used 32 million yen from the funds for supplies as operating funds, made lump-sum purchases of materials, and stored them. This is because if materials had been purchased separately using funds for materials included in the relevant items (construction expenditure, improvement expenditure, and operating costs) in accordance with a single year's budget, it would not only have caused problems for projects in the following year due to shortages of materials, but also would have hindered efficient use of the materials. The Ministry supplemented expenses for materials used in projects by using funds from the relevant items (construction expenditure, improvement expenditure, and operating costs) and managed flows from these items under the supplies account. Even though it could store materials, the Ministry could not store materials worth 100–170 million yen per year for long periods of time because it had to keep the balance for the stored materials and the factory account (for timber preservation treatment and the manufacturing of uniforms etc.) to 32 million yen or below (Hirayama and Fujikawa 1936). Also, it could sign multiple-year contracts up to 5 years into the future, but sleepers were not suitable for long-term storage due to potential cracking and rotting. It was also possible to change expenses for material purchases with the transfer of funds between budgetary items of the category *moku* or below, which was allowed to some degree; however, the highest priority was placed on coal purchases.[8] Therefore, it seems that the budget for purchasing other materials including sleepers was cut significantly. In sum, the only measure that the Ministry of Railways could take to deal with the lack of budget for material purchases was a reduction in contract prices.

7.2   Implementation of an Invited Tendering System

The Ministry of Railways implemented an invited tendering system for a limited period, from 1930 through 1932 (purchases for 1931 through 1933). This appears to be a measure the Ministry took to overcome the two aforementioned problems, namely, the shortage of timber suitable for sleepers and the tight budget. In other words, by implementing an invited tendering system, the Ministry tried not only to reduce contract prices and secure the necessary quantity of sleepers within the reduced budget, but also to secure sleepers with desirable quality through re-selection of designated sleeper dealers. In the invited tendering held in November 1930, the Ministry signed purchase contracts for a total of 3.03 million sleepers with 330 sleeper dealers out of the 360 dealers who were qualified to the tendering. Although details of the tendering are unknown, Yoshimoto Limited Partnership Co., which had been unable to get a contract since 1921, was among the 330 sleeper

---

[8]The share of railway sleeper purchases in the entire expenditures on materials was 6.3 % in 1929 and 5.3 % in 1931. In contrast, the share of coal increased from 26.4 % in 1929 to 37.8 % in 1931.

dealers which signed a contract. The company was finally permitted to supply 30,000 sleepers, but at the lowest price. The company purchased mountains, forests, and trees in Iwate, Fukushima, Nagano, and Gunma prefectures and produced sleepers in the second half of the 1920s; the transaction volume reached about 400,000 sleepers by 1931 (Tetsudō Jihō, 15 November 1930; Yui 1961). It seems that the Ministry of Railways tried not only to reduce contract prices, but also to add major sleeper dealers, which had grown as large as or larger than the existing designated sleeper dealers, as new members of the designated sleeper dealers.

Following implementation of the invited tendering system, the contract price of a sleeper fell from 2.06 yen in 1924 to 1.48 yen in 1930 and then to 0.88 yen in 1932. Similarly, the expenditure on sleeper purchases declined from 9 million yen in 1924 to 7.9 million yen in 1930, then to 4.85 million yen in 1932. This enabled the Ministry of Railways to handle the budget problem (Tetsudōshō 1920–1930). The Ministry could not, however, effectively deal with shortages of timber suitable for sleepers through the re-selection of sleeper dealers and, in addition, faced a further decline in the quality of sleepers. In the recessionary period, sleeper dealers took contracts at low prices in order to secure their sales channels. But if they could not meet orders with their stock of sleepers, they could do nothing but deliver cracked or bent sleepers. A number of sleeper dealers incurred losses and there were many contract violations in 1932 (Nihon Makuragi Kyōkai 1965). The shortage of timber suitable for sleepers might have been mitigated or eliminated by increasing the number of designated sleeper dealers; however, it probably was not easy to further increase the number of designated dealers (which had already been more than 300) and conduct contract negotiations repeatedly. Also, the decreased quantity of sleepers contracted per dealer in the 1930s, when the Ministry of Railways' demand for sleepers was satiated, decreased the dealers' incentive to supply the Ministry with sleepers.

7.3   Increased Demand for Timber, and Measures to Secure Sleepers

From 1933 onwards, as industrial activities were invigorated under the so-called Takahashi fiscal policy, the demand for timber increased, and after the Sino-Japanese War broke out in 1937 the demand for timber for military purposes soared. The price for general-purpose timber started to rise, and since profits in the Ministry of Railways' budgetary account returned to normal, the Ministry reverted to the traditional transaction process and tried to secure sleepers of guaranteed quality. The shortage of timber suitable for sleepers was still not mitigated at all.[9] Therefore, putting emphasis on improving rot-resistant timber, the Ministry of Railways

---

[9]The share of chestnut and hinoki wood in the sleepers purchased by the Ministry of Railways in 1937 was only 3.7 % and 4.2 %, respectively (Manshū Chōsabu 1939).

devised new ways to dry sleepers and decided on a plan to inject more effective amounts of preservatives into sleepers in order to drastically reduce the quantity of sleepers procured to a minimum level and to cope with the devastated mountains and forests (*Tetsudō Jihō*, 1 August 1936). At the Accounting Department's timber preservation treatment factory, which moved in 1930 from Fukagawa to reclaimed land at Shibaura close to Shinagawa Station, the Ministry of Railways not only conducted experiments and research on timber preservation, but also produced more than 10 % of the treated sleepers and supplied them mainly to the railway management bureau in Tokyo (Nihon Kokuyū Tetsudō 1972).

According a survey conducted by the Ministry of Railways in January 1936, 45 % of the sleepers installed nationwide (27 million sleepers) were treated sleepers. Especially in the jurisdiction of the Tokyo railway management bureau, the percentage of installed treated sleepers reached as high as 93 % (Suzuki 1938; *Tetsudō Jihō*, 29 August 1936). However, despite the use of treated sleepers, the Ministry of Railways could not reduce its demand for sleepers. This is because the Ministry increased the use of pine timber, which had the shortest useful life (6.4 years) among the tree varieties used for treated sleepers. Moreover, supplying pine timber to the Ministry was advantageous to the dealers. For example, Kodate Timber Co. purchased pine timber, which was relatively easily available, as 'supplementary timber', because of a reduced amount of hiba trees logged in Aomori, and because of their rising prices which resulted from increased competition among buyers (Tetsudō Jihō, 3 October 1936; Kodate Mokuzai Kabushiki Kaisha 1933–1938). The share of pine sleepers in all treated sleepers purchased by the Ministry of Railways increased from 9 % (140,000 sleepers) in 1927 to 38 % (740,000 sleepers) in 1936 and then to 60 % (1.91 million sleepers) in 1939 (Nihon Makuragi Kyōkai 1959). Also, a decline in the number of large-diameter logs led to a reduction in the useful life of sleepers regardless of tree variety used. This exacerbated the problem of sleeper degradation further. From the second half of the 1930s, as the domestic timber market tightened, the Ministry of Railways continued to search for solutions to the problems of procuring sufficient supplies.

8 Conclusion

JGR responded to changes in supply and demand in the markets for timber and sleepers and to fluctuations in the budget for material purchases, and shifted the mechanism for procuring sleepers from competitive tendering to sole-source contracts, then to invited tendering before returning to sole-source contracts. Among these methods, the competitive tendering system enabled JGR to purchase sleepers at the lowest prices. However, since it was difficult to secure the necessary quantity of sleepers with this system, JGR shifted the procurement method to sole-source contracts. With sole-source contracts, JGR attempted not only to have the designated sleeper dealers, who had sufficient funds and credit, deliver sleepers with certainty by giving them various incentives, but also to secure the necessary quantity of sleepers within budget by concentrating the procurement process in response to the expansion of areas from which timber was procured. Put differently, JGR chose a procurement mechanism that prioritized securing the necessary quantity of sleepers as long as procurement was accomplished within budget, even though the purchase price was high relative to the price offered in the general timber market.

The designated sleeper dealers, however, were not specialized in producing and selling sleepers for JGR, and therefore responded sensitively to trends in demand in other timber markets. During economic booms when the price of general-purpose timber rose, the designated sleeper dealers sold timber in other timber markets where they could earn greater profits, rather than supply JGR with sleepers. In contrast, during the boom times JGR could not secure the necessary quantity of sleepers, being unable to overspend its budget. Procuring sleepers with sole-source contracts made it difficult to reduce contract prices due to curbed competition, and JGR faced difficulties securing the necessary quantity of sleepers once a budget deficit emerged. The response to this problem during a recessionary period was to implement the invited tendering system. However, since it would become difficult to secure the necessary quantity of sleepers once the economy started to grow, JGR shifted the procurement method back to sole-source contrasts after solving its budgetary problem.

Overall, it can be said that the changes to JGR's procurement of sleepers were made in order to secure necessary quantities of sleepers while facing budget constraints. However, despite such a flexible response, JGR could not secure sufficient sleepers in the long run. Although it took short-term measures against changes in demand for timber or in its budget that were caused by macroeconomic fluctuations, JGR could not take long-term measures that would take into account not only static budget constraints each year, but also future procurements of sleepers. The first reason why JGR could not secure stable supplies of sleepers in the long run was its procurement method, and the second reason, which is also important, was a collapse of the balance between supply and demand. Although timber is a renewable resource, the speed of tree regeneration cannot be artificially controlled. If demand for timber increased due to industrial development, supply shortages occurred, which could in turn constrain the development process. Analysis of the procurement and use of timber in the process of industrialization is an examination of measures taken by industry in order to deal with the limits of the natural environment and can potentially explain the relationship between industrial development and the natural environment, an issue that remains to be studied in economic history. This article sheds light on only a part of the field, but shows that the railway industry was not free from constraints imposed by the natural environment.

Primary Sources

Hasegawa Tokyo Shiten (1895–1910, 1914 and 1915) *Kakuki kessansho (Kessanhyō)* (Profit and loss statement of the Hasegawa Tokyo Branch), nos 3–18, 22, 24.

Teishinshō (1893–1902) *Teishinshō kōbunsho kikai buppin* (Documents of the Ministry of Communications Railway materials), vols 1–9.

References

Dajōkan Monjokyoku (Document Bureau, Great Council of State). 1891 and 1903. *Kanpō* (Government gazette), nos 2277 and 6084.

Dai-Nihon Sanrin Kaihō (Bulletin of the Japan Mountains and Forests Association), March 1920, no. 448.

Gifu Shinbun (Gifu Newspaper), 3 December 1995.

Hasegawa Mokuzai Kōgyō Kabushiki Kaisha (Hasegawa Timber Co.). 1967. *Hasegawa Kyōji Shōten hachijū-nenshi* (The eighty-year history of Hasegawa Kyōji Co.).

Hasemoku. 1988. *Hasegawake Mokuzai hyakunen-shi* (A hundred-year history of the Hasegawa Family).

Hirayama, Takashi, and Fujikawa, Fukue. 1936. *Tetsudō kaikei* (Railway accounting). Tokyo: Shunjūsha.

Hokkai Times, 28 December 1918.

Kamimura, Yoshio. 1935. *Makuragi kaizen no kyūmu* (Urgent need for improving railway sleepers). Tokyo: Tetsudōshō.

Karumichō (Karumi Town). 1975. *Karumichō-shi* (The history of Karumi Town).

Kobayashi, Masaaki. 1977. *Nihon no kōgyōka to kangyō haraisage* (The sale of the government enterprises and the industrialization of Japan). Tokyo: Tōyō Keizai Shinpōsha.

Kōbe Yūshin Nippō (Kōbe Yūshin Daily), 16 October 1918.

Kodate Mokuzai Kabushiki Kaisha (Kodate Timber Co.). 1933–1938. *Eigyō hōkokusho* (Business reports), no. 14–19.

Kōjunsha (ed.). 1911 and 1914. *Nihon shinshiroku* (Who's who in Japan), 15th and 18th ed.

Makino, Fumio. 1996. *Manekareta Purometeusu: Kindai Nihon no gijutsu hatten* (An invited Prometheus: Technological development in modern Japan). Tokyo: Fūkōsha.

Manshū Chōsabu (Research Department, the South Manchurian Railway Co.). 1939. *Nihon narabi Manshū ni okeru tetsudō makuragi jukyū jōkyō* (An overview of the supply and demand of railway sleepers in Japan and Manchuria).

Matsunami, Hidezane. 1924. *Meiji ringyō shiyō* (A history of Meiji forestry), part 2. Tokyo: Dai-Nihon Sanrinkai (The Japan Mountains and Forests Association).

Matsushita, Takaaki. 2004. *Kindai Nihon no tetsudō seisaku: 1890–1922* (Railway policy in modern Japan: 1890–1922). Tokyo: Nihon Keizai Hyōronsha.

Mitsui Bunko. 2004. *Mitsui Bussan shitenchō kaigi gijiroku* (Minutes of the meeting of branch-heads of Mitsui Bussan) (Reprint edn), vols. 11 and 12.

Mokuzai Hozonshi Hensan Iinkai (Editorial Committee of the History of Wood Preservation). 1985. *Mokuzai hozon no ayumi to tenbō* (History and prospects of wood preservation). Tokyo: Nihon Mokuzai Hozon Kyōkai (Japan Wood Preserving Association).

Nagoya Shōgyō Kaigisho (Nagoya Chamber of Commerce). 1914. *Nagoya Shōgyō Kaigisho geppō* (Monthly report of the Nagoya Chamber of Commerce), no. 83.

Naikaku Kanpōkyoku (Government Gazette Bureau, Cabinet). 1900. *Hōrei zensho* (Compendium of laws).

Naikaku Tōkeikyoku (Statistics Bureau) (ed.). 1896–1937. *(Dai-)Nihon Teikoku tōkei nenkan* (Statistical yearbook of the empire of Japan).

Nakamura, Naofumi. 1998. *Nihon testudōgyō no keisei: 1869–1894* (The formation of the Japanese railway industry, 1869–1894). Tokyo: Nihon Keizai Hyōronsha.

Nihon Kokuyū Tetsudō (Japan National Railways). 1969, 1971 and 1972. *Nihon Kokuyū Tetsudō hyakunen-shi* (A hundred-year history of Japan National Railways), vols. 1, 7 and 9.

Nihon Kōtsū Kyōkai (Japan Transportation Association) 1952. *Kokutetsu no kaiko* (Reminiscences of the Japan National Railways). Tokyo: Nihon Kokuyū Tetsudō.

Nihon Makuragi Kyōkai. 1959. *Makuragi* (Railway sleepers). Tokyo: Nihon Makuragi Kyōkai.

Nihon Makuragi Kyōkai (Japan Railway Sleepers Association). 1965. *Nijū-nen no ayumi* (Twenty-year history of the Association). Tokyo: Nihon Makuragi Kyōkai.

Nihon Tōkei Kyōkai (Japan Statistical Association) (ed.). 2006. *Nihon chōki tōkei sōran* (Historical statistics of Japan), vol. 2.

Nōshōmushō (Ministry of Agriculture and Commerce). 1907–1921. *Nōshōmu tōkeihyō* (Annual statistics of agriculture and commerce), 22nd to 38th ed.

Nōshōmushō (Ministry of Agriculture and Commerce). 1910. *Tetsudō Makuragi* (Railway sleepers).

Oikawa, Yoshinobu. 1992. *Sangyō kakumeiki no chiiki kōtsu to yusō* (The local transportation during the period of industrial revolution in Japan). Tokyo: Nihon Keizai Hyōronsha.

Ōkurashō (Ministry of Finance). 1902–1943. *(Dai-)Nihon Gaikoku Bōeki Nenpyō.*

Ōkurashō (Ministry of Finance). 1920–1925. *Gaikoku bōeki gairan* (Overview of foreign trade).

Ringyō Hattatsushi Kenkyūkai (Research Group of the History of Forestry). 1958. *Mitsui Bussan Kabushiki Kaisha mokuzai jigyō enkaku-shi* (History of the timber business of Mitsui & Co.). Tokyo: Ringyō Hattatsushi Kenkyūkai.

Sawai, Minoru. 1998. *Nihon tetsudō sharyō kōgyo-shi* (A history of the Japanese Railway vehicles industry). Tokyo: Nihon Keizai Hyōronsha.

Suzuki, Ichigorō. 1938. *Tetsudō makuragi jukyū jōkyō* (Supply and demand of railway sleepers). Tokyo: Nōshōmushō.

Taniguchi, Tadayoshi. 1998. Zairai sangyō to zairai nenryō (The impact of indigenous industrial development on fuel consumption). *Shakai Keizai Shigaku* (Socio-Economic History) 64(4): 61–86.

Teikoku Tetsudō Taikan Hensankyoku (Editorial Bureau for the Overview of the Imperial Railways) (ed.). 1984. *Teikoku Tetsudō Taikan* (Overview of the imperial railways) (reprinted ed.), vol. 3. Tokyo: Hara Shobō.

Teishinshō (Ministry of Communications). 1904–1905. *Teishinshō nenpō* (Annual report of the Ministry of Communications).

Tetsudō Jihō (The Railway Times), 19 November 1921, 18 January 1930, 15 November 1930, 1 August 1936, 29 August 1936, and 3 October 1936.

Tetsudō Sangyōkyoku (The Railway Industry Bureau). 1898–1905. *Tetsudō Sangyōkyoku nenpō* (Annual report of the Railway Industry Bureau).

Tetsudōchō (Railway Authority). 1906 and 1907. *Tetsudōchō nenpō* (Annual report of Tetsudochō).

Tetsudōin (Railway Authority). 1908–1920. *Tetsudōin nenpō* (Annual report of the Railway Authority).

Tetsudōin (Railway Authority). 1916–1919. *Tetsudōin tetsudō tōkei shiryō* (Statistics of the Railway Authority).

Tetsudōshō (Ministry of Railways). 1920–1943. *Tetsudō tōkei (shiryō)* (Statistics of the Ministry of Railways).

Tomiyama, Kiyonori. 1934. Tetsudō yōhin (Railway materials). In *Keiri kaikei yōhin chōsa* (Investigations of materials of the accounting office). Tokyo: Tetsudō Kenkyūsha.

Umemura, Mataji, et al. (eds.). 1966. *Nōringyō* (Agriculture and forestry) (*Chōki keizai tōkei* (Long-term economic statistics), vol. 9. Tokyo: Tōyō Keizai Shinpōsha.

Yamada, Hikoichi. 1911. Tetsudō makuragi ni tsuite (On railway sleepers). *Gifuken Sanrin Kaihō* (Bulletin of the Gifu Prefecture Mountains and Forests Association) no. 6.

Yamaguchi, Asuka. 2008. Senzenki Nihon no tankōgyō ni okeru kōboku chōtatsu (The role of timber in the prewar Japanese coal-mining industry). *Shakai Keizai Shigaku* 73(5): 23–46.

Yamaguchi, Asuka. 2009. Senzenki no Hokkaidō tankōgyō ni okeru kōboku chōtatsu (The role of timber in the prewar Hokkaidō coal-mining industry). *Mita Gakkai Zasshi* (Mita Journal of Economics) 102(2): 187–212.

Yokohama Bōeki Shinpō (Yokohama Trade News), 9 October 1909.

Yoshitsugu, Toshiji. 1951. *Kokutetsu no shizai* (Materials for Japan National Railways). Tokyo: Hitotsubashi Shobō.

Yui, Tsunehiko (ed.). 1961. *Yoshimoto Gojū-nen no ayumi* (Fifty year history of Yoshimoto Timber Co.). Tokyo: Yoshimoto Ringyō Kabushiki Kaisha/ Yoshimoto Gōshi Kaisha (Yoshimoto Limited Partnership Co.).

Chapter 3
The History of Ecological Environment: Ideas Derived from Chinese Research

Makoto Ueda

Abstract During the last few years the phrase "*seitai-kankyō*", the Japanese term for "ecological environment", has appeared frequently in the scholarship of Japanese researchers of China. The two words that make up this phrase were coined in Japan as translations for concepts imported from Europe, and then exported to China. Having passed through the filter of the Japanese language, their nuances are not necessarily the same as the terms used in the United States and Europe. In addition, the phrase "*seitai-kankyō*" itself is the Japanese re-import of the Chinese phrase, "*shengtai huanjing*", formed by joining the originally Japanese words "*seitai*" and "*kankyō*".

From the 1990s, problems of pollution, global warming, deforestation, desertification, and water scarcity surfaced in China. It became clear that unusual fluctuations in the level of the Yellow River were causing "manmade", rather than "natural", disasters. At this time, "ecology" was influenced by the field of biology, while "environment" was coloured by ideas from engineering. A term that would combined both ideas was needed leading to the appearance of "*shengtai huanjing*". When Chinese research on the ecological environment was finally introduced to Japan and joint research began, one result was the adoption of this Chinese term by Japanese researchers.

Keywords Ecology • Environment • Ecological environment • Ecosystem • Socio-ecological history

This is a translation of an article that originally appeared in Shakai Keizaishi Gakkai ed., *Shakai Keizai Shigaku no Kadai to Tenbō* (Issues and Prospects for Socio-Economic History). Tokyo, Yūhikaku, 2002, pp. 105–118.

M. Ueda (✉)
College of Arts, Rikkyō University, 3-34-1 Nishi-Ikebukuro, Toshima-ku, Tokyo, 171-8501, Japan
e-mail: ueda@rikkyo.ac.jp

1 Introduction

During the last few years the word "*seitai-kankyō*", the Japanese term for "ecological environment", has appeared frequently in the scholarship of Japanese researchers of China. The term is derived from two phrases, "ecology" (*seitai*) and "environment" (*kankyō*), both of which were coined in Japan for use as technical terms in translation after the base concepts were imported from Europe. However, the term "*shengtai huanjing*", the equivalent Chinese term for "ecological environment", was born in China, where the two phrases above were joined, and the term was then imported to Japan.

The term "ecological environment" differs in nuance from "natural environment", its Japanese counterpart. The meaning of the word "natural environment" can vary according to the user, yet its location is normally external to the subject to which it relates—i.e. humanity. Thus, the idea that humanity is a utilizer must be included in its identity. However, "nature" is also generated "spontaneously" without being constructed by humans, and so the surroundings become the environment (Torigoe 2001) when focus is placed upon humans as the center of discussion. The term natural environment separates humans from nature, and yet it also forces us to recognize the mutual relationship between both entities. On the other hand, the "ecological environment" does not alienate nature from humankind, and identifies both as players within a single system. Rather, scholars analyzing the relationships instead observe the effects of human behavior within the system.

When observing long-term social change as a subject of historical research, it is clear that humans have meddled with nature at numerous points in time. The resulting environment becomes a secondary "nature" that regulates human behavior during later stages of the historical process. By focusing upon this process as an object of research, rather than prioritizing the framework of recognition defined as the natural environment, it becomes more feasible to apply systems theory to the analytical framework of the ecological environment. This change in methodology serves as the backdrop for the use of the term ecological environment in Japan.

Turning to the matter of origins, what was the reason for combining the two phrases of "ecology" and "environment" in China? What kind of framework of recognition was used in this combined term? In this article, the route by which the term transformed will be traced from the past into the present. Afterwards, China's ecological environment will be surveyed. Finally, the article argues that the methodology centered upon examining the ecological environment should be rechristened as ecological history.

2 Ecology and Environment

The term "*seitaigaku*", the Japanese term for "ecology" was originally coined by Miyoshi Manabu (1861–1939), the father of Japanese botany, who founded the field after returning to Japan from a period of study abroad in Germany during the Meiji era. The word was initially created for the purpose of translation, and first appeared when Miyoshi translated the German "Pflanzenbiologie" into "plant ecology" within his Japanese book under the title of *Recent Progress in the Field of European Biology* published in 1895 (Kimura 1993). There is thus a need to explore how "*seitaigaku*" was linked to the concept of "biology" rather than "ecology".

What becomes clear upon tracing Miyoshi's footprints is that the movement that he founded was not intended to end as simple botany. While studying in Germany, Miyoshi gained a profound interest in the natural preservation movement. This influence led him to submit to the House of Peers in 1911 the "Proposal related to the preservation of historic relics and natural monuments", in which he dedicated his efforts toward the establishment of the Natural Monument Preservation Law. Similarly, in 1929, when construction began to widen a road through the virgin forests of Nara Prefecture's Mount Kasuga for the purpose of facilitating tourism, Miyoshi conducted a field survey after local botanists requested an estimate of the amount of damage likely caused to the flora. He went on to point out the influence of increased travelers upon the mountain in terms of the change in vegetation type and the damage from exhaust fumes. As a result of this report delivered to the Ministry of Education, the construction companies in Nara prefecture were ordered to halt their work (http://osaka.yomiuri.co.jp/home.htm 2001). The project was adjusted so that the road was instead fashioned from gravel, and although in the end construction still was forced into the area, this is an example of a pioneering effort in the movement to preserve natural scenery.

The nuances of Miyoshi's constructed word *seitaigaku* cannot be adequately reflected in the term botany. Rather, an awareness of ecological conservation and the recognition of humanity's impact upon vegetation are also included. In other words, the term ecology is not a simple translation of the German "biologie". Instead, the term fits both within the rubric of academia, where it acts as part of the methodology of vegetation surveys, and also simultaneously serves as a foundation for the preservation movement of nature.

In terms of the ecosystems of forests or tidal plains, these locations are not recognized as legal entities within Japan's present day judicial system. For this reason, specialists in ecological research are trying to clarify the value of ecology, and are frequently participating within the preservation movement. The majority of Japanese ecologists are attempting to answer the demand for these services. The character of the academic field opened by Miyoshi has continued into today in this way.

On the other hand, the original term "ecology" itself was coined by the German zoologist Ernst Haeckel (1834–1919) in 1866, when he added the suffix of –logy to the Greek word *oikos* (meaning "family finances" and "domestic economy") to emphasize the "academic nature" of the studies. In terms of etymology, its origins are essentially the same as the term "economy". Following the First World War, Haeckel's principles of biological origination were thought to have served as an inspiration for Nazi theories of eugenics, and thus some individuals are hesitant to use his term "ecology" (Iijima 2000).

During the 1980s, the Green Party gained significant support as interest in ecological dangers increasingly became shared by the intellectual community. In this group's discourse, we can observe careful efforts to avoid mixing any Nazi ideologies into their political platform. On the other hand, in other sections of Europe many believe that ecology and Nazism share certain affinities. For this reason, there is a trend toward preferential use of the term "environment" over "ecology".

In the United States, the term "ecology" has different connotations than in Europe. At the end of the nineteenth century, Ellen Swallow Richards (1842–1911), a genders studies specialist at the Massachusetts Institute of Technology, attempted to illustrate how research about air, water, and soil quality could be used to observe the destruction of the environment caused by the American industrialization. Arguing that the environment could be improved through efforts within the home and family, she called the study of environment and life "ecology". This was in 1892. Interestingly, Swallow Richards was not a biologist by training, but a chemist. Swallow Richards is also called the "mother of home economy" (Robert Clarke 1973). Swallow Richards' term of "ecology" was also a term that originated from the Greek *oikos*, but differed from Haeckel's version. The American version of "ecology" could be considered synonymous with the Japanese *katakana* word.

Miyoshi's *ecology* is not synonymous with the German "Okologie". It is also unrelated to Swallow Richards' term. It is distinct from the above described versions of "ecology" that must be considered within wider political contexts. Thus, Miyoshi's "ecology" and the "ecology" of other groups are not equivalent. In the Japanese version of "ecology", the concept is rooted within the framework of biology, yet also is linked to social issues. Because the word "ecological" entered China after the Cultural Revolution, shades of the Japanese version of the term colour its meaning.

Turning now to the second phrase that makes up the term "*seitai-kankyō*", the word "environment" came into frequent use in Japan during the 1970s when pollution became a societal problem. At that time, the environment was considered separate from human society. Thus, the environment became the location where humans discharged pollutants like mercury, sulfur oxides, and nitrogen oxides. Environmental problems were seen as solvable as long as human interference was limited, and regulations regarding different kinds of waste products were legislated. The government organ set up to oversee and manage these issues in Japan was the Environmental Agency. Thus, the word environment emerged in tandem with the idea of public pollution.

After this time, the word "environment" became a tool with which to defend the actual environment, and it thus spread. Iijima Nobuko has skillfully summarized the change in the use of the term, and thus it will not be described further here. Iijima bases her argument upon a summary of documentary history, but also succeeds in defining the Japanese word as "an entity that encapsulates the entirety of all external conditions and influences necessary for the existence of all living creatures. If human society is thought to be the central subject with which it interacts, the environment is thus a concept with both socio-cultural and natural-scientific-physical dimensions" (Iijima 2000, p. 2). This remains the most recent definition of "environment". The current circumstances surrounding the idea differ greatly from those of the 1970s.

China's equivalent for "environment", the word "*huanjing*", was imported in 1973 during the first National Conference on the Preservation of the Environment. At this conference, the idea of "total-use plans" related to "three concurrent activities" (equipment for the prevention of pollution would be integrated during the design, building, and utilization phases of building construction) and "three pollutants" (waste water, waste gas, and solid waste products) was affirmed. In other words, the context in which the "environment" was discussed in China mirrored that of the discourse related to the "environment" of 1970s Japan, but subsequent developments in the environmental sciences went in separate directions. In China, the word "environment" is discussed primarily in terms of regulating pollutants. Instead of discussing the idea simply as one's physical environment, it is more akin to an abbreviation of the concept "environmental protection". Thus, problems surrounding ecological issues like the destruction of forests or desertification are apt to be omitted from the super-category of the "environment".

In summary, the terms "ecology" and "environment" in Chinese cannot be directly translated into English as "ecology" and "environment", because they have passed through a filter called the Japanese language. Thus, because of the various historical conditions surrounding its context in Japan and the different routes of intellectual circulation in China, the terms are enfolded in nuances different than those in the United States and Europe.

When the 1990s began, problems limited not only to pollution, but also global warming, deforestation, desertification, and the loss of sources of fresh water surfaced in China. At this time, "ecology" was influenced by the field of biology, while "environment" was coloured by ideas from engineering. Thus, a new term which combined the two ideas was needed. It was because of this era that the Chinese version of the term "ecological environment" emerged.

3   China's Crisis in the "Ecological Environment"

Beginning in the 1980s, intellectuals from outside China began to increasingly point out the developing crisis in its ecological environment. One of the comparatively early and comprehensive studies on the issue was written by Vaclav Smil (1983). Smil based his survey upon empirical data to demonstrate the direct dangers to China caused by acute deforestation and desertification, the loss and pollution of sources of fresh water, air pollution, and the reduction of bio-diversity. In Japan, similar pioneering efforts were made by Fukao Yōko (1990, 1992).

Starting in the 1990s, the state policy of rapid economic development stifled efforts towards democratization, and also inflated the crisis in the ecological environment. Beginning in 1995, the annual study called the "State of the World", published by a group led by Lester R. Brown named World Watch, pointed out the huge, negative impact that China's economic development was having upon the global environment (Brown et al. 1995). Brown also explored this theme in further single-author publications (Brown 1995). Based upon the empirical data, Brown sounded a warning about the dangerous conditions in China stemming from the rapidly increasing demand for food, which would soon surpass the limits of the ecological environment to provide such resources. He also found that similar trends existed across the world.

Because the warning came first from foreigners, China initially denied the criticisms. However, Brown's cautionary tale about the massive global burden from China's increasing droughts and demands for cereal garnered global attention. The Chinese state refuted the argument on the basis that Brown's data was insufficient. However, within the country itself, there was wide-spread recognition that a failure to avoid a crisis in the ecological environment threatened the basis of economic growth. China's economic policy was modeled upon Japan's high economic growth process. Japan's frequent struggles with pollution-related illnesses like the Minamata Mercury Poisoning and the Yokkaichi Asthma helped create an awareness of problems that needed to be avoided.

In advanced countries like Germany, the United States, and Japan, economic development was initiated prior to the development of environmental problems, so support for ecological problems or the creation of environmental movements opposed the ideology of economic development as espoused by modernization theory. However, since China was able to learn from the experiences of the advanced countries, the conservation of the ecological environment was instead designated as a prerequisite for modernization (called the "creation of the present" in China). There exists a belief in Japan that one can have "only" the economy or ecology. However, both concepts are seen as parts of a greater whole in China, and regardless of the current situation, such a harmonious balance was seen as a given.

Both "eco-" words emphasize these aspects of harmony, but the concept that most encapsulates both terms is the idea of "ecological economics" proposed in China in 1980. According to Fukao, "If one was to define the idea of 'ecological economics', it is the concept that in order to have positive development in the future, one must be aware of ecological conditions. Furthermore, one must understand the path of development within a particular region and minimize the burden upon the environment as much as possible. In so doing, one establishes a system of material cycling that is comparatively independent within the region" (Fukao 1990).

More concretely, one example is the use of methane gas emitted as waste from villages as a fuel. Today, there is advanced testing of this kind of zero-emissions policy. The author visited such a project being designed at Yunnan Province's Ecological Economics Academic Society in 1992. The area is part of a valley far from the coast, so the region receives little rainfall and desertification was proceeding rapidly. In accordance with the plan, a mid-sized dam was built between two mountains, and water from the lake behind the dam was used to irrigate the vegetation on the mountains. Soil erosion was controlled in order to extend the lifetime of the dam by creating fruit orchards for crops like grapes. Thus, the regional economy was supported while conservation of the ecological environment was also promoted. What most vividly left an impression upon the author was the Chinese word for dam, "ecological water storehouse".

Just as can be seen in the example of this dam, Japanese "ecology" and Chinese "ecology" differ in their aims. Among Japanese activists who participate in the "ecology" movement, the resistance towards building dams is quite high. The idea of constructing a dam for the purpose of conserving the ecological environment would be unheard of. This is one of the differences in these types of ecological environments between China and developed countries.

This situation only began to change between 1997 and 1998. In 1997, the water level of China's Yellow River dramatically fell, and a situation called "Yellow river shear flow" developed in which river water failed to reach the ocean. During that year the final 700 km length of the river dried up for a period of 7 months. The following year, the Yangtze river and the Northeastern Songhua river both flooded. Victims of the disasters climbed to 220 million people.

The shear flow and flood were not simply natural disasters brought about by weather change. Awareness spread that deforestation and desertification in the upper and middle portions of the rivers were closely linked to the shear flow and flood, leading many Chinese people to turn their gaze towards the ecological environment. Furthermore, an acknowledgement of the need to suppress the prioritization of pursuing profits in order to conserve the ecological environment finally appeared.

During the same time period, the third United Nations Framework Convention on Climate Change (UNFCCC) was convened in Kyoto in December of 1997. Capitalizing upon this opportunity, many Japanese corporations that had previously supported economic growth came to sing the praises of conserving the ecological environment. The idea of pursuing both economic development and environmental conservation became common sense, and the assumption that modernization opposed environmental protection began to fade.

In short, the recognition that global environmental problems were a shared issue for humankind broadly diffused at the end of the twentieth century. Differences in awareness between Japan and China disappeared thanks to this trend. Furthermore, Chinese research on the ecological environment finally was introduced to Japan, and joint research began to be conducted. The use of the Chinese term "ecological environment" by Japanese researchers was one such result of the process.

4 Research Conditions in the Study of the Ecological Environment

In terms of historical research in the field, an awareness by Chinese scholars of the historic fluctuation in the ecological environment is felt even more keenly than by those in Japan. When Hara resided in Beijing in 1999, she wrote that "I came to understand that Chinese historical research on the environment was not only very broad, but deep in its understandings as well. Naturally, the methodologies used were moving in valuable directions that were hard to overlook for Japanese research—no, the totality of research—on Chinese environmental history" (Hara 2000). While reevaluating and reworking past research it became clear that the topic was one that extended all the way back into classical history, thereby demonstrating that the crises in the ecological environment were not just problems that recently appeared due to modernization.

There is unfortunately insufficient space to comprehensively chronicle the historiography within this article, and the author's own resent documentary surveys of the material remain incomplete. Thus, the author would like to select representative studies that characterize the field from a number of research areas central to the history of the ecological environment.

The relationship between the ecological environment and human society has always been a central theme in the field of historic geography. Here, two representative researchers of the field will be introduced. Chen Qiaogi examines the regions of Shaoxing and Ningbo in Zhejian province, while Shi Nianhai studies the Loess Plateau spreading across Northwest China. Chen's work examines the development of lake and marsh regions, along with deforestation. Meanwhile, Shi consistently studies the transformations in the ecological environment in the Loess Plateau. His results are summarized in *Heshanji* (Works on Rivers and Mountains) (Shi 1963, 1981, 1988, 1991a, 1991b, 1997, 1999).

Shi relies primarily on documentary sources such as official histories and literature, and collects the fragmentary data related to the ecological environment. In this way, he has surveyed the change from the classical period through to the Ming and Qing eras, in order to create a foundational survey of empirical data. The attractiveness of this research is that he has presented a grand hypothesis based upon the clues previously ignored by other researchers, such as modification of the names of rivers and the frequency of flooding, and changes to the names of vegetation.

In one of Shi's examples, he describes how the scenery of the Loess Plateau prior to the Warring states era was completely different than the current version. The mountains were cloaked in forests, grasslands covered the plains, and an eroded valley had not yet formed. Reclamation of the land began during the Warring states era, and large sections of the plateau were cultivated during the early Han dynasty. It was here that erosion became extreme and the Yellow river became muddy from the soil. The basis of this hypothesis comes from tracing the name "Yellow river". This name first appeared during the early Han dynasty, and prior to that it had simply been called "river". Naturally, one of the reasons for the establishment of this new name was the expansion of the Chinese world, and the inclusion of other great rivers necessitated the modification of older names. Because of this political context, previous research did not link the change of names to a transformation in the ecological environment. Regardless of the veracity of this hypothesis, we can see Shi's research style in this example.

Shi's research only began to garner attention in Japan in the 1990s. In particular, Tsuruma Kazuyuki, who researches the Qin and Han dynasties, and Seo Tatsuhiko, who studies the urban history of the Tang (Seo 2000), are both central to the effort to further develop Shi's conclusions. Additionally, studies in this area have increased following the creation of a joint research program between the Shaanxi Normal University Historical Geography Research Center, where Shi serves as Center Director, and Japanese scholars (the MEXT research grant "Research on the History of Castle towns on China's Loess Zone and the Ecological Environment", 1997–1999).

One portion of the results of the joint research was published in a special edition of *Asia Yūgaku*, and later released in a separate publication (Ueda 2002). It was encouraging to see interest in the ecological environment blossom for those young Japanese researchers who participated in the joint research. However, how future research endeavors will influence the field is as of yet undetermined.

Approaching similar questions from a different vantage point than historical geography are the efforts of He Yeheng, who carefully collected the source materials that documented change in the environmental ecology over the long duree (He 1996 The data is mainly gathered from regional records that mention the infestation of animals and plants, and is very useful as foundational research. However, only a scant 500 volumes were printed, so it is regrettable that the number of researchers who will be able to see these records will be limited outside of those who focus on China. Tackling a different theme, Robert B. Marks centers his research on the tiger (Marks 1998), as does Ueda Makoto (1998c, 1999b). The history of the tiger's relationship with humankind will continue to be a fascinating theme, and the methods used to investigate the topic still remain to be settled.

Historical research is also being deepened by fieldwork in ethnology and cultural anthropology. One representative of this trend is Yin Shaoting. In his research, instances of slash-and-burn agriculture are located within documentary sources, where he investigates the problem of whether there is change in this primitive form of agriculture. He combines fieldwork with documentary analysis to prove that there has been change in this behavior over time (Yin 1994, 1996, 2000a, b, 2001).

For example, there is the popular theory that agriculture "matures" from slash-and-burn farming on mountain lands to wet-paddy cultivation. However, in contrast to this idea, one can find Neolithic ruins along old shorelines, and in documentary records scholars observe that the Yue people living in Southern China were skilled in travelling along river ways. Similarly, in the present day many ethnic groups that still practice slash-and-burn farming are descendants of migrants that were forced to relocate during the Ming and Qing eras, thus throwing the older periodization into question. Furthermore, if such details as the type of crop grown, the order plants were cultivated, or the tools used are examined, one observes that such cultivation methods were not the remains of primitive farming styles. Rather, Yin's research empirically demonstrates that it was a form of farming deliberately selected by the people who lived in that area, in order to find a style that fit the ecological environment. Additionally, regarding slash-and-burn agriculture, he later authored a number of studies where more documents were identified that reinforce this idea.

Yin's research also serves as a criticism of the theory of East Asian evergreen forest culture, which was once popular in Japan. On account of this theory, the culture of ethnic minorities such as those observed in Yunnan had historically been ignored and their origins were assumed to be similar to those of peoples found in Japanese cultures. However, for these cultures that had developed since ancient times in the ecological extremes of certain regions, if it can be assumed they changed over time, they would need to be analyzed from the perspective of the members of the subject culture itself. In the end, the level of research in the field increased by leaps and bounds thanks to the research illuminating the connection between the ecological environment and culture of the people of Yunnan.

Last, it is critical to also introduce the work of Zhao Gang (1996) who has succinctly, but in a well-balanced way, chronicled the change in China's ecological environment. He focuses on topics like the relationship between human migration, development, and the ecological environment; the relationship between forestry policy and land rights; the process of degradation in the ecological environment as seen in the Loess Plateau and desert regions of northwestern China; the links between the Qing dynasty policy of reclamation and the shed people; the historic trends in the consumption of forest resources; the influence of the expansion of cultivated land following the burial of rivers and swamps; and the frequency of natural disasters caused by the degradation of the ecological environment. It is based upon evaluations of each of these topics that he has bound together his hypotheses.

Thanks to the research results in the field of historical geography conducted by scholars like Shi Nianhai and Chen Qiaoyi, such as the examination of the influence of the shed people on the ecological environment in their mountain-region home, scholars have added significantly to the historic research on China. This evolution is symbolized by the increase in contact between scholars thanks to the accumulated research on the shed people in Japan (Ueda 1994, 1997, 1998a; Shibuya 2000). One hopes that research exchanges investigating historical change in the ecological environment will be furthered through a translation of such material into Japanese.

5   Future Plans in the Field of Ecological History

Historical research related to China's ecological environment has surged in Japan in recent years. In the studies of Miyazaki Yōichi (1994) and Oka Motoshi (1998) among others, efforts have been made to carefully collect documentation on the field. Also, the work to introduce Chinese research as a whole is also increasing (Kubo 2001). However, as for researchers who analyze the theme of historic change in the ecological environment, only Hara Motoko, in the field of classical history, and Ueda, in the field of Ming and Qing history, have placed the topic at the core of their own research. What Hara and Ueda share is a methodology in which they utilize knowledge of natural sciences when reading the historical source material.

In Hara's *magna opus* on the development and the environment in classical China (Hara 1994), she utilizes data from the *Guan-zi* (Writings of Master Guan) local edition. In the first half of this text, the names of soil types are cited, and in the second half the species of crops and fruits that can be produced in a particular order are recorded. Hara uses soil science, agriculture, and plant biology when viewing the relationship between the environment and people during the classical age. In this study, the scientific Latin names are always included when describing the names of plants. When incorporating knowledge of both animal and plant biology into historical research, the scientific names of the objects serve as a useful reference.

Ueda labels his own research as ecological history, and is trying to establish a new research sub-field. In the 1980s, Ueda focused his main research on the relationship between ethnic groups and local society as seen through the lens of genealogy. However, he changed his research focus to the ecological environment after encountering data in the genealogical texts listing times of death, which caused him to become interested in analyzing changes in the rates of death depending upon the season (Ueda 1988). In order to understand societal changes over a duration exceeding a hundred years in length, he focused not only upon relationships of cause and effect, but also incorporated ideas of positive feedback within a framework of systems theory. This became the start of his work where he proposed the paradigm of historic systems.

The paradigm of historical systems is a framework that seeks to understand historical event in terms of the three dimensions of ecological systems, social systems, and semantic systems. The methodology analyzes trends in the flow of matter, human relationships, and information, as all systems possess these characteristics. Ecological systems and systems of meaning have each been defined previously in separate studies (Ueda 1996, 1998b). Additionally, the meaning of ecological environment was posited based upon clues in theories of *feng shui* (Ueda 1999c).

Methodologies of analyzing ecological history are located within the broader field of plant sociology (Ueda 1999a). The study of plant sociology is a sub-field in the field of plant ecology, established by Braun-Blanquet. In this sub-field, communities of plants, which are defined as a collection of plants, are viewed as a society composed of multiple species. The goal of the field is to survey the species that comprise the community and clarify the mutual relationships between the plants. Plants grow in an area not by accident, but instead are entwined in relationships with other plants. When humans meddle in the world of natural vegetation, change occurs in the patterns of growth. These processes of change progress in a particular order, and the transformation of plant society, in other words the vegetation there comprises the community, can vary on this level.

By utilizing this knowledge of plant sociology to confirm the existence of certain species of plants in particular conditions, it becomes possible to estimate the degree and character of human interference. If one pursues this line of thinking, scholars should be able to investigate the state of past plant communities and the ways in which humans interfered with those societies by using the vegetation whose existence can be identified within the historical sources.

This kind of technique does not need to be limited to plant sociology. For example, if one knows that a particular type of animal existed in the past, one can apply the knowledge of animal ecology and locate the societal links in terms of what foods were consumed or estimate the area in which the animal lived by reconstructing its ecological environment. Furthermore, if one utilizes one's knowledge of epidemiology, one can use historic facts related to when a particular disease circulated to estimate the routes of infection, the habitat or ecological environment of the era, and the existence of animals that might have carried the illness. The author investigated the transmission of plague as an example of this kind of research (Ueda 2000, 2001).

In the evidential school of China, the work of identifying the names of animals, plants, and diseases that are located within the historical sources is steadily progressing. However, the work of the evidential school ends at the stage where one says, "In essence, A is B". In comparison, ecological history uses ecology as a guide, and aims to recognize the existential order within the ecological environment. Thus, the field continues to evolve (Ueda 2002).

6 Conclusion: Socio-economic History and Ecological History

Sociology in the broad sense (studies focused upon human, plant, and animal societies) is a field that seeks to identify the formation of order between multiple elements. If that is a viable definition, ecological history should also be able to be referred to as socio-ecological history. The sounds of the terms may be slightly jarring, but here the stem eco- is the axel of the concept, and points to a multilayered relationship with socio-economic history.

During the 2001 conference of the Society of Socio-Economic History, the issue of environmental history was centrally debated. A framework for ecological history was posited by Ueda in the discussion panels, where he argued that the "condition" for creating capitalism was derived from the ecological environment. Similarly, he proposed that the "external environment" of economic activities should be methodologically analyzed side-by-side with the ecology. However, despite the shared interest regarding the question of how to handle the history of the ecological environment, a common awareness of the issue did not emerge between the different presenters.

To use the definition of economics provided by Paul Samuelson, "Economics is the study of how men and society choose, with or without the use of money, to employ scarce productive resource which could have alternative uses, to produce various commodities over time and distribute them for consumption, now and in the future among various people and groups of society." (Samuelson 1970). Current economics focuses upon the valuation of non-physical goods such as services and intellectual property rights, so Samuelson's definition has become outdated. However, his definition still proves useful in situations where economics are used to study developments in the past. On the other hand, ecology can be defined as "research with the goal to analyze the relationship between living species or individuals of those species that are influenced by the flow of energy and matter".

When defining and comparing both concepts, economics is a field that examines the flow of material from the perspective of humankind; ecology, in comparison, is a field that analyzes the flow of material from the perspective of all living creatures except humankind. A salient feature of economics is the ability to investigate the flow of material that lies outside the boundaries of ecosystems. There are two ways in which this process manifests. One is "trade". Throughout history, humankind has engaged in large-scale trade that has exceeded the boundaries of the local ecology, for example, in terms of the exchange of silk woven in temperate regions with horses reared in arid regions. This kind of trade also has the potential for transforming ecosystems. The other form is "mining". Fossil fuels and ores can be dug from the depths of the earth where few living creatures interact. Humans can acquire the power to change ecosystems through this process. When trade and mining are used to interact with ecosystems, we can see the core elements of ecological history.

Just as global environmental problems have become an aspect of the field of ecology, so too is this the case in economics. Thus, socio-economic history and ecological history have become the flip-sides of the same coin, and are the elements to forming a new field of inquiry.

References

Brown, R. Lester. 1995. *Who will feed China?* New York: Norton.

Brown, R. Lester, et al. 1995. *State of the word 1995.* New York: Norton.

Clarke, Robert. 1973. *Ellen Swallow: The woman who founded ecology.* Chicago: Follett Pub. Co.

Fukao, Yōko. 1990. Chūgoku ni okeru seitaiteki kiki to mirai (China's ecological crisis and the effect on its future). In *Gendai Chūgoku no teiryū* (Undercurrents of China today), ed. Mitsuru Hashimoto, and Yōko Fukao, Tokyo: Kōrosha.

Fukao, Yōko. 1992. Chūgoku ni okeru shisutemuronteki seitaigaku kenkyū (A system theory approach to ecological studies of China). *Ajiagaku Ronsō* (Journal of Asian Studies) 2:143–156.

Hara Motoko. 1994. *Kodai Chūgoku no kaihatsu to kankyō* (Development and the environment in ancient China). Tokyo: Kenbun Shuppan.

Hara Motoko. 2000. Sakkon no Chūgoku ni okeru kankyōshi kenkyū no jōkyō (A survey of recent studies in the environmental history of China). *Chūgoku Kenkyū Geppō* (Monthly Report of the Institute of Chinese Affairs) 628:42–46.

He Yeheng. 1996. *Zhongguo hu yu Zhongguo Xiong de Lishi Bianqian* (A historical examination of tigers and bears in China). Changsha: Hunan Normal University Press.

Iijima, Nobuko. 2000. *Kankyō mondai no shakaishi* (A social history of environmental problems in Japan). Tokyo: Yūhikaku.

Iijima, Nobuko. 2001. Kankyō shakaigaku no seiritsu to hatten (The formation and development of environmental sociology). In *Kōza kankyō shakaigaku* (Studies in environmental sociology), 1, eds. Nobuko Iijima, et al. Tokyo: Yūhikaku.

Kimura, Makoto. 1993. Miyoshi Manabu. In *Seitai no jiten* (The dictionary of ecology), ed. Makoto Numata. Tokyo: Tōkyōdō.

Kubo, Takuya. 2001. Chūgoku kankyō hogo shiwa yakuchū (5) (Translator's notes to stories about the history of the protection of the environment in China). *Fukuyama Daigaku Ningen Bunka Gakubu Kiyō* (Journal of the Faculty of Human Culture and Sciences, Fukuyama University) 1:19–60.

Marks, Robert B. 1998. *Tigers, rice and silt: Environment and economy in late imperial South China*. Cambridge: Cambridge University Press.

Miyazaki, Yōichi. 1994. Min-Shin jidai shinrin shigen seisaku no suii (Changes in forest resource policies in Ming and Quin China). *Kyūshū Daigaku Tōyōshi Ronshū* (The Oriental Studies, Kyūshū University) no. 22.

Oka, Motoshi. 1998. Nansōki setsutō kaikō toshi no teitai to shinrin kankyō (The economic stagnation of the Zhedong Seaport Cities and deterioration in the forest environment in the Southern Song Period). *Shigaku Kenkyū* (Historical Studies, Hiroshima University) no. 220.

Samuelson, Paul A. 1970. *Economics*, 8th ed. New York: McGraw-Hill.

Seo, Tatsuhiko. 2000. Kankyō no rekishigaku (History studies of environment). *Ajia Yūgaku* (Intriguing Asia) no. 20.

Shi, Nianhai. 1963. *Heshanji* (Works on rivers and mountains) 1. Beijing: Joint Publishing Co.

Shi, Nianhai. 1981. *Heshanji 2*. Beijing: Joint Publishing Co.

Shi, Nianhai. 1988. *Heshanji 3*. Beijing: Beijing People's Press.

Shi, Nianhai. 1991a. *Heshanji 4*. Xi'an: Shanxi Normal University Press.

Shi, Nianhai. 1991b. *Heshanji 5*. Taiyuan: Shanxi People's Press.

Shi, Nianhai. 1997. *Heshanji 6*. Taiyuan: Shanxi People's Press.

Shi, Nianhai. 1999. *Heshanji 7*. Xi'an: Shanxi People's Press.

Shi, Nianhai. 2000. Kan-tōdai no chōanjō to seitai kankyō (Chang'an castle and the ecological environment during the Han and Tang Periods). *Ajia Yūgaku* no. 20.

Shibuya, Yūko. 2000. Shindai Kishū Kyūneiken ni okeru hōminzō (The shed people in Xiuning Prefecture, Huizhou Province, during the Qing Period). In *Dentō Chūgoku no chiikizō* (Images of the regions in traditional China), ed. Eishi Yamamoto. Tokyo: Keiō Gijuku Daigaku Shuppankai.

Smil, Vaclav. 1983. *The bad earth: Environmental degradation in China*. New York: M. E. Sharp Inc.

Torigoe, Hiroyuki. 2001. Ningen ni totte no shizen (Nature from a human point of view). In *Kōza kankyō shakaigaku* (Studies in environmental sociology), 3, ed. Hiroyuki Torigoe. Tokyo: Yūhikaku.

Ueda, Makoto. 1988. Min-Shinki Setsutō ni okeru seikatsu junkan (Life cycles in eastern Zhejiang Province during the Min and Qing Period). *Shakai Keizai Shigaku* (Socio-Economic History) 54(2):64–91.

Ueda, Makoto. 1994. Chūgoku ni okeru seitai shisutemu to sanku keizai (Ecological systems and the mountainous economies of areas in China). In *Ajia kara kangaeru* (Asia as a starting point), 6, ed. Yuzō Mizuguchi et al. Tokyo: Tokyo Daigaku Shuppankai.

Ueda, Makoto. 1996. Shiteki shisutemuron to busshitsuryū (Historical system theory and patterns of distribution). In *Shichō* (Current of History) 38.

Ueda, Makoto. 1997. Sanrin oyobi sōzoku to kyōyaku (Regulations involving clans and forests). In *Chiiki no sekaishi* (World history from a regional perspective), 10, ed. Seiji Kimura, et al. Tokyo: Yamakawa Shuppannsha.

Ueda, Makoto. 1998a. 'Sanrin kenzoku' to shinrin hogo (Rights to forested mountains and forest protection). *Gendai Chūgoku Kenkyū* (Modern and Contemporary China Studies) 2.

Ueda, Makoto. 1998b. Shiteki shisutemuron to jōhōryū (Historical system theory and patterns of information distribution). In *Chūgoku minshūshi e no shiza* (Chinese history from the viewpoint of the common people), ed. Kanagawa Daigaku Chūgokugo Gakkai. Tokyo: Tōhō Shoten.

Ueda, Makoto. 1998c. Tora no me kara mita chiiki kaihatsushi (The history of regional development in China from a viewpoint of tigers). In *Iwanami kōza kaihatsu to bunka* (Iwanami studies in development and culture), 5, ed. Yōnosuke Hara. Tokyo: Iwanami Shoten.

Ueda, Makoto. 1999a. *Mori to midori no Chūgokushi* (An ecological history of China). Tokyo: Iwanami Shoten.

Ueda, Makoto. 1999b. Zōkibayashi o meguru tora to hito (Tigers, people and the use of woods). *Chūgoku: Shakai to Bunka* no. 14.

Ueda, Makoto. 1999c. Kannō suru daichi: Fūsui (Chinese geomancy: The response from the earth). In *Kōza ningen to kankyō* (Studies in people and the environment), 10, ed. Masataka Suzuki. Tokyo: Shōwadō.

Ueda, Makoto. 2000. Kanteisho, pesuto to mura (Expert reports on villages and the plague). In *Sabakareru saikinsen* (A tribunal on bacteriological warfare). ed. *731 Saikinsen Saiban Kyanpen Iinkai and ABC Kikaku* (The Committee of the Campaign for a Tribunal on Bacteriological Warfare Research of Unite 731 and the ABC Agency), Tokyo: Saikinsen Saiban Kyanpen Iinkai and ABC Kikaku.

Ueda, Makoto. 2001. Saikin heiki to sonraku shakai (Bacteriological weapons and village society). In *Shippei, kaihatsu, teikoku iryō* (Disease, development and imperialist medical treatment), ed. Masatoshi Miichi, et al. Tokyo: Tokyo Daigaku Shuppannkai.

Ueda, Makoto. 2002. Seitaigakuteki rekishigaku no hōhō (Methods in ecological history). In *Chūgoku kōdo kōgen no kankyō to rekishi* (An environmental history of the huangtu plateau in China), ed. Kazuyuki Tsuruma, and Tatsuhiko Seo. Tokyo: Tōsui Shobō.

Yin, Shaoting. 1994. *Senlin Yunyu de Nonggeng Wenhua* (Forests and agriculture). Kunming: Yunnan Renmin Chubanshe.

Yin, Shaoting. 1996. *Yunnan Nonggeng Wenhua de Qiyuan: Nonggeng Juan* (Agriculture in Yunnan), 2 vols. Kunming: Yunnan Jiaoyu Chubanshe.

Yin, Shaoting. 2000a. *Ren yu Senlin* (People and forests). Kunming: Yunnan Jiaoyu Chubanshe.

Yin, Shaoting. 2000b. *Unnan no Yakihata* (Slash-and-burn farming in Yunnan) (translated by Shirasaka Shigeru). Tokyo: Nōrin Tōkei Kyōkai.

Yin, Shaoting. 2001. *People and forests: Yunnan Swidden agriculture in human-ecological perspective*. Kunmin: Yunnan Education Publishing House.

Zhao, Gang. 1996. *Zhongguo Lishi shang Shengtai Huanjing zhi Bianqian* (A history of the ecological environment in China). Beijing: Zhongguo Huanjing Kexue Chubanshe.

Chapter 4
The Problem of Air Pollution During the Industrial Revolution: A Reconsideration of the Enactment of the Smoke Nuisance Abatement Act of 1821

Masahiko Akatsu

Abstract The Industrial Revolution in Britain led to widespread pollution in the form of factory smoke, and raised the issue of social relief. Scholars have argued that the Smoke Nuisance Abatement Act of 1821 resulted from the efforts of just one public-spirited politician. In this paper, however, through examining parliamentary debates on this issue, we analyze how politicians, landlords, and industrialists viewed the damages caused by air pollution, how they developed a framework for redress, and how they interwove their interests into the act.

Landowners, and even the manufacturers who were responsible, recognized that air pollution caused damage to property. For this reason, the act was promoted by landowners and even some industrial capitalists, although they are normally regarded as opponents of smoke regulation. As smoke prevention techniques of the time might be a source of profit for manufacturers, the act of 1821 did not conflict with their business principles. The owners of private property were merely seeking redress for damages that they had suffered. It was not until 1840s that more interventionist legislation aimed at helping urban labourers was contemplated.

Keywords The industrial revolution • Air pollution • Public nuisance • Coal smoke • Urbanization

This is a translation of an article that originally appeared in Shakai Keizai Shigaku 69(4) (January 2003), pp. 71–91.

M. Akatsu (✉)
School of Political Science and Economics, Meiji University, 1-1 Kanda-Surugadai, Chiyoda-ku, Tokyo, 101-8301, Japan
e-mail: akatsu@meiji.ac.jp

© Socio-Economic History Society, Japan 2015
S. Sugiyama (ed.), *Economic History of Energy and Environment*, Monograph Series of the Socio-Economic History Society, Japan, DOI 10.1007/978-4-431-55507-0_4

1 Introduction

It is no exaggeration to say that concerns over environmental issues are mounting day by day. For its 2001 Annual Conference, Socio-Economic History Society of Japan chose as its theme, "Toward an Economic History of the Environment: Forests, Development, and Markets", which generated animated discussion about how natural resources and the environment should be addressed from the vantage point of economic history.

This article examines as economic history the air pollution caused by furnace discharges of coal smoke during the Industrial Revolution. It was during those years that air pollution, still a major environmental concern today, first became a significant social problem throughout Britain.

Mass consumption of coal in the Industrial Revolution began in Britain. The output of coal smoke increased, and with it the damage it caused. The British Parliament accordingly recognized that coal smoke was a public nuisance inflicting widespread damage, and in 1819, established a "Select Committee on Steam Engines and Furnaces" (hereinafter referred to as the Committee), to investigate smoke pollution and abatement measures. In 1821, Parliament went on to enact an "Act for giving greater Facility in Prosecution and Abatement of Nuisance arising from Furnaces used and [sic] in the working of Steam Engines" (Source 7, 1 and 2 George IV, c.41) (hereinafter referred to as the Act). This was probably the first anti-air pollution act to be applied to the whole of Britain.

The Industrial Revolution era was thus a critical period in the history of air pollution problems in Britain. Nevertheless, it has not yet been subjected to sufficient examination. There are those Japanese economic historians who argue that the Industrial Revolution marked the beginning of environmental problems, and who therefore regard pollution as a critical area for research on the Industrial Revolution, but the detailed research itself has yet to be carried out. Yoshiaki Takei cited many areas requiring research by economic historians, including public health, urbanization, and pollution, but by way of example could point only to the existence of a report drawn up by the Committee (Takei 1984, p. 217). Similarly, Yoshiyuki Sekiguchi and Jun'ichi Umetsu said, "The Industrial Revolution was the historical starting point of an era groping for ways to enable continuous economic growth to coexist with environmental preservation", but they only mentioned the Act's enactment as a basis for their argument, without offering any details (Sekiguchi and Umetsu 1995, p. 116). There are also jurists in Japan who have raised this issue, but they, like the economic historians, only refer to the Committee and the Act without giving any details (Tamura 1965, p. 64; Sugai 1974, p. 89). Thus, although many Japanese scholars date both the social problem of air pollution and the attendant abatement measures to the Industrial Revolution era, detailed research remains to be done.

What is the case in other countries? Carlos Flick's "The movement for smoke abatement in nineteenth century Britain" (Flick 1980), and Eric Ashby and Mary Anderson, in *The Politics of Clean Air* (Ashby and Anderson 1981), observe that the Industrial Revolution was an important period in the history of air pollution in Britain, and tried to address the issue based on parliamentary papers. These researches marshaled only superficial facts, however, without offering sufficient analysis. (1) Since there is little examination of the recognition of damage (especially economic) to victims, the status and motives of those advocating abatement not necessarily clear. (2) Since factory owners, though important people in their own time, are dismissed as mere polluters, their interest in pollution issues and their roles in anti-pollution measures are not given enough consideration. (3) The fact that technical measures not only benefited the public through smoke prevention but also garnered profit for factory owners is overlooked, and hence, (4) the progress that was made in anti-pollution measures is attributed solely to the goodwill, public spirit, or political aptitude of a given bill's proponent, while (5) the delay in abatement measures is reflexively attributed either to factory owners' general resistance or to technical difficulties.

This article is based on an examination of the parliamentary papers that have already been read by earlier scholars and to analyze them from the viewpoint of economic history, being especially cognizant of (a) the interests and motives of those involved, including factory owners, (b) the reconciliation of public benefit with private profit, and (c) the economic background of the legislation.

2 Coal Smoke as a Public Nuisance

2.1 The Rapid Expansion of Coal Consumption During the Industrial Revolution

Needless to say, it was not in the Industrial Revolution that coal smoke was discharged for the first time. It is well known that people in Britain, who began to use coal earlier than those in other European countries, were exposed to coal smoke continuously from medieval times (Gimpell 1975; Te Brake 1975; Brimblecombe 1988; Ōba 1979). However, it was during the Industrial Revolution that air pollution became a nationwide concern, because until then the problem was limited mainly to London where an exceptionally large amount of coal was consumed. Nationwide coal consumption in Britain increased about six-fold during the Industrial Revolution (from 1750 to 1830), and the ratio of coal consumption in London to the total decreased from about 16 % to about 9 % (Mitchell 1988, pp. 244–245, 247). In other words, coal consumption was rising steadily in areas other than London.

The decisive factor in this quantitative and geographic expansion was the shift from agriculture to industry, and the rapid improvement in productivity caused by the spread of machines and the factory system. Since most industries relied on heat treatment processes, rapid industrial growth caused coal consumption to grow wherever industry was located. The cotton industry, for example, which was located in Lancashire, increased its output about 70 times (More 1997, p. 47) and simultaneously increased its coal consumption, because of the many heat treatment processes, including bleaching and dyeing, that are used in textiles. At the same time, coal consumption also rose drastically in the iron industry in South Staffordshire and South Wales, which had converted from charcoal-to coal- or coke-use.

Early in the eighteenth century, steam engines were used on a limited basis use for drainage and pumping, but Watt's improvements made them useful for machines, and when his patent expired in 1800, coal became the dominant source of power. Neither formal statistics nor estimates exist showing the aggregate use of steam engines in early nineteenth century Britain, but if the estimates of the number of steam engines as of 1800, when their spread began, is supposed to be accurate, there were 2,191 steam engines in total. Of these, 1,064 were used in mines, 469 in the textile industry, 263 in the metallurgy (including iron and iron manufacturing), and 112 in the food industry including the brewing industry and distillers (Kanefsky and Robey 1980, pp. 169, 181). And their use was widely distributed over Britain: 271 in Yorkshire, 266 in Lancashire, 173 in Cornwall, 156 in Shropshire, 150 in Staffordshire, 139 in Northumberland, 136 in London, and 110 in Durham (Kanefsky and Robey 1980, pp. 176–177). Given that total steam power increased rapidly from 20,000 hp in 1800 to 300,000 hp in 1850 (Musson 1976, pp. 422, 435), it can be surmised that the number of steam engines had already increased considerably by 1820 when Parliament began to debate the issue of smoke pollution.

However, the expansion of coal consumption is not the only reason that air pollution became a nationwide problem during the Industrial Revolution. Urbanization in northern and western industrial areas, the concentration of factories and works, the agglomerations of dense populations around factories, and an increase in the number of sufferers, aggravated the problem of smoke.[1] As is often pointed out, the spread of steam power played an important role in advancing urbanization, by doing away with industry's dependence on hydraulic power and therefore with the restrictions that nature had imposed on industrial location (Landes 1977, p. 3). As the use of steam engines spread to the developing industrial cities, the number of people who could potentially suffer damage from coal smoke increased accordingly.

2.2 From District to State

2.2.1 Local Judgments of Coal Smoke as a "Nuisance"

As mentioned, there had been complaints about the damage caused by coal smoke ever since medieval times. However, until the Industrial Revolution, these

---

[1]For the increase in population of major cities, towns, and boroughs, refer to (Sweet 1999, pp. 3–4).

complaints came mainly from members of the exclusive classes, such as the royal family, the aristocracy, and courtiers, in London. As a result, smoke prevention measures were aimed at private relief for aristocrats rather than at public relief. This appears characteristically in a bill brought before Parliament and rejected in 1623, which sought to "prohibit... the burning of seacoal in brewhouses within a mile of any building in which the King's court, or the court of the Prince of Wales, should be held, or in any street west London Bridge" (Nef 1966, p. 157). From so local and personal an issue, however, coal-smoke pollution became a national social problem during the Industrial Revolution. What had been detested by aristocrats in London came to be regarded as an 'evil', a 'nuisance', and a 'public nuisance' by people throughout Britain. At last, in 1805 and 1812, inhabitants brought complaints against factory owners for the 'public nuisance' of discharging quantities of smoke and steam.

I would like here to explain the terms, 'public nuisance' and 'nuisance'. A 'public nuisance' in Britain is a type of nuisance that is illegal under common law, and it is distinct from a 'private nuisance'. Although the definition of a 'nuisance' changes somewhat according to the times and the interpreters involved, it is generally considered to be an infringement on another person's land or property, and to include the diffusion of certain toxic substances to the land, property, or body of another person that that person finds unpleasant or troubling. A 'nuisance' is judged to be 'private' or 'public' depending on the number of sufferers and the nature of the damage it inflicts. In the case of an act that injures neighbors' peace and comfort, such as discharging smoke or emitting bad smells, the number of sufferers is treated as an important standard for judgment. In the case of smoke, the discharge of black smoke is judged to be a 'public nuisance' not only when it damages others' health, but also when it damages neighbors' occupation or property (Katō 1979, pp. 269–75).

Based on early nineteenth century complaints of smoke as a public nuisance and the definition of public nuisance, it is clear that air pollution by coal smoke no longer remained a specific individual's problem but had become a community problem during the Industrial Revolution. Although the lawsuit in 1805 against Davy, a coke manufacturer, was lost (Rex v. Davey and Another, 5 Espinasse 217, 170 English Report 791), the plaintiffs in an 1812 lawsuit against Dewsnap, a steam engine owner in Sheffield, won their case (King v. Dewsnap, 16 East 194, 104 English Report 1063). It was confirmed legally that discharging smoke could be judged a public nuisance.

It was to be expected that many other complaints about coal smoke would follow, so local authorities began to consider abatement measures. In around 1800, the Police Commissioners of Manchester established a subcommittee on smoke nuisances and discussed measures to address them (Webb 1963, p. 259; Horibe 1975, p. 63). As a result, the Manchester Police Act of 1806 (or 1808) made it lawful for police commissioners to force a party (or parties) to reduce the discharge of smoke, where a complaint of nuisance had been brought by the neighbors under common law, and to fine the party (or parties) 2 lb a week if the party (or parties) failed to adopt the required abatement measures within a certain period.

Subsequently, about 20 lawsuits concerning smoke nuisances were brought in Manchester by 1820 (Source 1, p. 16). In 1816, a similar local Act was also enacted in Birmingham, a central city in the West Midlands that was known as 'Black Country' because of the smoke there (Sweet 1999, p. 89).

2.2.2 The First Step in Parliament

The many lawsuits involving smoke nuisances and the newly adopted measures by local authorities at last moved Parliament to act in 1818, when Member of Parliament Michael Angelo Taylor petitioned Parliament for the establishment of a select committee to investigate smoke nuisances. The Committee, under the chairmanship of Taylor, held its first public hearing on 14 June 1819.

As everyone knows, the early nineteenth century in Britain marked the starting point of the development of a liberal capitalistic economy. Parliament's economic policy was therefore also beginning to take on a liberal coloring. At the same time, it was just starting to regulate economic activities in order to address certain social problems, such as child labor (for example, the Factory Act of 1802 and 1819), gradually adopting interventionist policies of the kind seen in the so-called "age of reform" from the 1830s to the 1870s.[2] Like labour problems, pollution problems are mainly the result of economic activity, and any investigation into air pollution problems or institution of anti-pollution measures will inevitably mean some kind of intervention in the economy.

However, the early nineteenth century was also a period when the interests of industries using coal began to have a degree of influence that could not be disregarded in policymaking. The Parliamentary Reform Movement had already been launched in the second half of the eighteenth century, in an effort to make the landlord-dominated Parliament reflect industrial interests as well. It gained momentum going into the nineteenth century, as the percentage of the population that was concerned with industry rose (Nakamura 1961, pp. 27–33). Given these circumstances, Parliament could not intervene in the production of industries using coal or steam without taking their interests into consideration. The Committee was established against that backdrop.

3   The Committee and Its Character: Landlords and Manufacturers

The authority which Parliament referred to the Committee was "to inquire how far it may be practicable to compel Persons using Steam Engines and Furnaces in their different Works, to erect them in a manner less prejudicial to Public health and

---

[2]For the parallel progress of laissez-faire policies and interventional policies in the nineteenth century Britain, refer to (Muraoka 1980, pp. 248–254; Okada 1987, pp. 146–180).

Public comfort; and to report their Observation thereupon to the House" (Source 2, p. 3). In other words, the main duty of the Committee was to investigate the possibility of forcing the owners of steam engines and furnaces to adopt some means for addressing the nuisance of smoke, and to examine the various possible means in order to determine which could guarantee the desired result. It is not generally known that, as this assignment shows, the means for abatement already existed even before the Committee was established. One of the pioneering developers of smoke abatement was in fact James Watt, who acquired a patent for a design to diminish smoke in 1785. Ward, Thompson, Roberton, De Prony and others had also acquired related patents between the latter eighteenth and early nineteenth centuries (Source 3, p. 93). The above-mentioned Manchester Police Act was enacted in the context of the existence of these inventions. Therefore, the purpose of the technical inquiry by the Committee was not only to verify the existence of these measures but also to confirm that they were effective and cost-appropriate enough to be imposed on factory owners.

The Committee was made up of Members of Parliament Taylor, Charles Mills, Dugdale Stratford Dugdale, Kirkman Finlay, and Henry Monteith. These were the investigators into smoke nuisances and the anti-smoke policymakers of the Industrial Revolution era.

We can guess at Taylor's status, given his title of 'Esquire'. Roughly speaking, Britain at the time was divided into the landed class (lords and the gentry), the bourgeoisie (industrial capitalists, merchants, bankers, etc.), and the working class. Men with the title of 'Esquire' were the gentry, mainly minor local landlords. They held public office, such as mayor, alderman, and justice of the peace, and played a central role in local governance. At the same time, they also held many seats in the House of Commons. They were what is called the 'Squirearchy' (Nakamura 1976, p. 11; Cannon 1997, pp. 405–406). Although Taylor was MP for the city of Durham, he changed his constituency frequently, and therefore did not necessarily represent the interests of that city or the industries located there. According to Gerrit P. Judd's classical study on the profile of the MP, Taylor was a mere landlord having no occupation and no relation with commercial or industrial interests (Judd 1955, p. 351).[3] Taylor's direct motive for his petition in 1818 and for launching the Committee was damage from smoke that he himself suffered: the flowers planted in the yard of his residence in White Hall, London, were blackened by smoke from the steam engine of the Lambeth Waterworks located on the opposite shore of the Thames, and the smoke also hindered walking in the yard. Taylor invited Robert Banks Jenkinson, Second Earl of Liverpool, who was a neighbor and the Prime Minister, to join him in the complaint (Source 4, 18 April 1821, col. 440; Source 1, p. 10). In other words, the first proponent of abatement measures was a landlord unrelated to industry, and a sufferer from coal smoke. He was not just a reformer with a rich sense of public duty as earlier studies have shown, but a person with a

---

[3]Taylor is also known for his eager concern about the improvement of streets in the City of London. For details, refer to (Webb 1963, pp. 290–294).

clear interest in the issue of smoke nuisance. Mills, MP for the borough of Warwick, and Dugdale, MP for Warwickshire, were likewise landlords (Judd 1955, pp. 179, 277).

Not only landlords like Taylor, but manufacturers, too, were Committee members. In 1819, some years before the passage of the 1832 Reform Act, people with strong ties to manufacturing industries had only 13 seats out of the 658 in Parliament, and they had only 9 seats in 1820 (Judd 1955, p. 89).[4] Monteith and Finlay were examples of Members with industrial connections. Monteith was MP for the burghs of Linlithgow, Provost of Glasgow, and also a calico manufacturer with a factory that had installed two steam engines. Although it is not known what type of industry he engaged in (probably textiles), Finlay, MP for Malmesbury, was also a manufacturer (Judd 1955, pp. 70, 193, 280). Their participation in the Committee does not seem to have been happenstance. It is clear that the Committee was strongly conscious of the interests of industry in its considerations. Neither were Monteith and Finlay unwilling watchdogs in judging on whether or not to adopt Taylor's measures. Monteith was a claimant for abatement as well as Taylor. He said at a public hearing that smoke was "certainly a very public nuisance about Glasgow" and agreed to participate because taking some measures would be also "a benefit to myself" (Source 1, p. 11).

Numerous witnesses were summoned to the Committee's public hearings. Since the main purpose of the Committee was to investigate means of abating smoke, the developers of plans, design, apparatuses, and methods constituted the first group of witnesses and were largest in number: William Moult of Whitby, Joseph Gregson of Liverpool, John Walker of London, William Losh of Newcastle, William Brunton of Birmingham, John Wakefield of Manchester, and Josiah Parkes of Warwick, etc. Most were civil engineers specializing in technical matters. However, among them was a manufacturer, Parkes, who produced worsted with his family. This shows that manufacturers who might contribute to smoke pollution were actively seeking the means to prevent smoke.

The second witness group consisted of manufacturers summoned to share their opinions about the commercial use at their factories of the methods and apparatuses presented by the developers: Birmingham metal worker William Phipson, London brewer Frederick Perkins, London soap-boiler Benjamin Hawes, London distiller James Scott Smith, and Peter Whitfield Brancker, a sugar refiner in Liverpool, etc. Importantly, the reason they were chosen from among many manufacturers was that they had independently and willingly adopted means to prevent smoke at their factories before the Committee was even established.

Finally, two doctors were summoned to testify about the damage to health caused by smoke. These were physician Edward Roberts and doctor Leman Tuthill, both of London.

---

[4]However, many manufacturers may have been included among the members classified as 'a merchant', 28 in 1819, and 30 in 1820. Although the member Calvert who appears in section 5 is classified as 'a merchant', he is considered to be a manufacturer.

The Committee itself was established by Taylor, the landlord who suffered damage to his garden caused by smoke. Under the "Squirearchy" of 1820s Britain, it was natural that a landlord with a personal complaint would take the lead against smoke nuisances. Unexpectedly, however, many manufacturers also participated in the Committee as representatives, from the brewing distillery industries and from the textile cotton and woolen industries. The Committee might have rejected their participation, as they were the leading consumers of coal and steam power, but given the socio-economic circumstances of the time, their interests could not be disregarded, as their presence on the Committee attests. Moreover, Monteith's testimony in public hearings also suggests that factory owners cooperated in Taylor's action of their own accord.

4 Recognition of Smoke-Damage and Motives for Preventing It

How did the British in the early nineteenth century become aware of the damage caused by smoke? As discussed, Taylor became interested in smoke nuisances and launched the Committee because of the damage to his garden. His recognition of the damage was directly linked with his motive for action, and proved very important. This motivation has been disregarded in previous studies, however; even those that mention Taylor treat his as a unique case. Here, I will address how people other than Taylor became conscious of the damage, and will classify their concerns as either damage to health or economic damage. Economic damage is especially important when we consider the position of the factory owner and his role in the battle against smoke.

4.1 Damage to Health

Two doctors on the Committee asserted that smoke pollution was damaging to health. One of them, referring to a contemporaneous report comparing longevity in London and in the countryside, stated that the duration of life was "considerably diminished by a residence in this metropolis", and said "whatever we can discover in such atmosphere, so as to render it different from that of the country, may contribute to produce this effect; and therefore it becomes probable that the quantity of carbonaceous matter suspended in it, is one of the causes of its insalubrity", and that "The rapid advancement to recovery which we frequently see in sick persons, during a short residence in the country, prove the influence which the atmosphere of London has upon health." Acknowledging that many factors other than smoke might shorten life, he nevertheless claimed that the "violent tornados of smoke" emitted by brew houses, soap manufactories, gas houses, and etc. was significantly harmful to health (Source 2, pp. 10–11).

Why did they believe that smoke injured health and shortened the lifespan? Another doctor answered that, "air so loaded with impurities, cannot be so respirable as is necessary to the health of the population. From this cause, the air loses its elasticity, and cannot be properly ventilated." He also said that smoke "vitiates the pure air, and mechanically deprives it of the power of ventilation" and that "it deprives the air of its vital principle, and produces an unwholesome atmosphere." This represented the medical knowledge of early nineteenth century Britain regarding the impact of smoke on the human body. It would have been easy to claim that smoke damages health, if it could be proved that coal smoke directly caused one or another specific illness. But both doctors denied such a link. When the Committee asked one of them, "Do you think that smoke has a tendency to generate particular disorders?", the doctor responded clearly, "I do not." Moreover, the other doctor stated that smoke does not cause respiratory diseases like asthma, and that only those who have already suffered from respiratory ailments are tormented by smoke (Source 2, pp. 8–11). The level of chemical knowledge of coal smoke at the time strongly influenced the doctors' replies.

Today, ash dust, SOx, NOx, etc. are considered to be the main ingredients of the smoke that results from the combustion of a fossil fuel. However, as the doctors' testimonies reveal, people at the beginning of the nineteenth century believed them to be mainly ash dust, carbon, and soot. If a person inhales ash dust in large quantities over an extended period, he will suffer respiratory ailments like the pneumoconiosis or asthma from which miners suffer. Such misfortunes did not readily befall ordinary town residents, however. The contaminants of smoke that directly cause specific diseases are gaseous substances such as SOx and NOx. These are far more harmful than ash dust, and cause bronchitis, pneumonia, and asthma (Kurachi 1982, pp. 76–81). Needless to say, smoke contained these toxic substances in the nineteenth century as it does today, but as they were not fully understood, a direct connection could not be made between smoke and illness. Chemical and medical knowledge was limited to the vague recognition that smoke must somehow adversely affect health. This was a significant feature of early nineteenth century Britain's understanding of smoke-damage, and also a critical limitation.

4.2 Economic Damage

While chemical and medical knowledge was ambiguous about the damage smoke caused to health, economic damage had a rather more significant meaning for the progress of abatement measures. Many witnesses described this kind of damage, pointing to the depreciation of buildings, the negative impact on products and factories, the contamination of houses and gardens, and so on. Gregson, who was a real estate agent and an engineer,claimed that a building of his had suffered a loss in value. He said he owned two or more buildings in Liverpool that he leased, but that as a result of a factory's being established in the neighborhood and beginning to emit much smoke in around 1813, his buildings, which had been worth 400 lb a year, had fallen in value to 100 lb a year (Source 2, p. 5). This was Gregson's motive for developing methods for abating smoke. Smoke damage led also to a fall in the price of land, which explains the motives of the real estate agents and landlords who demanded abatement measures.

Monteith, a calico manufacturer, and Parkes, a worsted manufacturer, asserted that smoke had a bad influence on their products or factories. Parkes introduced a new bleaching process developed in around 1814 into his factory in Warwick, but suffered damage to his bleaching spaces from the soot emitted by an old steam engine in his factory 10 yards from that space (Source 1, p. 5). He developed an apparatus of his own to diminish smoke, in order to avoid its damage. Moreover, Monteith, who acknowledged that a particular instance of smoke nuisance in Glasgow was indeed a "public" nuisance, also was angry that his factory and products were tormented by smoke, and therefore consented to join Taylor's efforts. Those engaged in the food industry, too, suffered the same kind of damage. A brewery in Salford introduced a means for purification (Source 1, pp. 11, 15–16), which suggests the existence of damage from soot to products or materials in a brewery. Air pollution might inflict damage not only on landlords but also on manufacturers themselves. This fact prompted manufacturers to take part in the development of anti-pollution policies.

Moreover, the problem of house- or garden-contamination did not affect only such landlords as Taylor. An alum manufacturer in Whitby adopted anti-smoke processes in one of his two factories in order to prevent damage to his house caused by the smoke emitted from that factory (Source 2, p. 9). Since a large number of factory owners still lived near their own factories in those days, their houses and property were at risk of smoke damage.

Previous studies tended to regard Taylor as the only active reformer and promoter of anti-smoke policies. However, it is clear that factory owners also recognized the damage from which they themselves suffered, and that they, too, played a significant role in developing anti-pollution measures.

5 Smoke-Abatement Measures and Factory Owners

5.1 Means for Eliminating Smoke

Needless to say, the question of whether the technical means in fact exist matters in the effort to battle air pollution. The only smoke-abatement measures that can be effective without the requisite technical methods are to prohibit the use of coal or to force workplaces to move. These were tried before the Industrial Revolution.

But forcing such measures on industries using coal during the Industrial Revolution would inevitably have led to their stagnation. A crucial question, then, was how to prevent smoke without checking the development of industry. The very fact that technical means existed at the time marks the Industrial Revolution as the starting point of "coexistence" between economic development and environmental protection.

The Committee was shown two types of technical approaches based on a common principle. One was an advanced furnace developed by Gregson, Wakefield, Parkes and others. It was designed to increase air inflows into the furnace with a port or pipe (ports or pipes) set beside the fire grate, which was the original source of air inflow. Another type was a revolving grate, developed by Bunton. It was designed to increase air inflows by gently rotating the fire grate (by means of a steam engine) and flattening the coal on it uniformly. The principle underlying both types was the same. With increased air (oxygen) inflows into a furnace, coal would be burned completely and the soot (carbon) that was believed to be a main component of smoke would be consumed.

These technical improvements were originally developed for furnaces that heat boilers for steam engines. However, the Committee was investigating not only furnaces for boilers but also general furnaces, including blast furnaces, glass furnaces, kilns for porcelain manufacture, and pots for brewing or distillation. The Committee therefore asked one witness about his attempts to apply these new means to general furnaces. He testified that, where high temperatures were required, as in the case of blast furnaces, abatement measures rendered the temperature too low, whereas for pots for brewing, distillation or soap-boiling, or for furnaces for the fumigation of brewing barrels, which did not need high temperatures, these technical approaches could work (Source 2, pp. 5–6). An experiment was actually made by Parkes on a brewing pot and fumigation furnace at a brewery in London. Some difficulties caused by the irregular nature of thermal demand were found in the brewing pot, but the experiment was largely successful with regard to the fumigation furnace (Source 1, pp. 9, 17). This was, however, only an experiment. The majority of Committee members recognized that practical experience with general furnaces remained insufficient. Therefore, the legal measures ultimately adopted by Taylor were not applied to furnaces at large, but were limited to furnaces for steam engines.

In any case, the Committee confirmed the existence of technical means of removing smoke from steam furnaces. That was not its only charge, however. The other and more important obligation of the Committee was to confirm that this means was effectual and cost-effective enough to be imposed on factory owners. And this is an important matter for us too, as we seek to understand the problem of air pollution during the Industrial Revolution from the viewpoint not of technological history but of economic history.

5.2 Costs and Effect

5.2.1 On Initial Costs

Advanced furnaces did not need special equipment or machinery, and even existing furnaces were able to use the new abatement means with only trivial or partial repairs. Costs, therefore, were not high. For the furnace developed by Gregson, they were 16 lb per boiler for a 6 hp steam engine, and 30–32 lb per boiler for a 36 hp engine; in the Wakefield furnace case, about 16 lb per boiler for a 20 hp steam engine, and for Parkes's furnace, 20–30 lb per boiler. And since repair was easy, it could be completed in 2 or 3 days, so that losses from an idled plant were minimal. Therefore, the advanced furnaces had a good reputation among manufacturers. For example, one brewer who adopted Parkes's furnace and paid 30 guineas (31 lb and 10 shillings) per boiler, testified that "I was shocked by the low price and the simplicity"(Source 1, p. 8). As the total capital of his brewery in 1830 was 758,000 lb (Musson 1978, p. 135), 30 guineas would be trifling for him indeed. The revolving grate, on the other hand, was a complicated machine, and costs for its installation were about 300 lb per boiler for a 20 hp steam engine. Because of the difference in cost, the Committee hoped to be able to endorse the advanced furnace over the revolving grate. However, as will be seen, the problem of initial costs ultimately became a less important consideration for the Committee.

5.2.2 On Performance

How much effect did the technical means of that time have on smoke abatement? Needless to say, the developers of the proposed means all said that they were effective. What was the testimony of the manufacturers who actually used these means in their factories?

It is important to note that both advanced furnaces and revolving grates were already in use by many manufacturers, and that both were highly regarded. A Birmingham metal worker who owned two steam engines (100 hp and 60 hp) and consumed 60 t of coal per week, visited Parkes's factory before the installation of his new furnace. His observation was that "he [Parkes] has attained complete success in the consumption of smoke." Moreover, one brewer, who asked Parkes to repair his two furnaces, said that Parkes's apparatus "answers most completely, as far as regards the destruction of smoke", and a brewing craftsman who worked under that brewer also testified that "since the adoption of Mr. Parkes's plan there is scarcely any smoke after the first of the morning". Furthermore, a soap boiler in London, who asked Parkes to repair a furnace, also said, "We find a very considerable reduction of smoke in consequence", while a distiller, who adopted Brunton's expensive equipment, testified, "We have found that we can consume the smoke to a very great extent, and although it is not completely invisible, yet it is never offensive" (Source 1, pp. 7–9, 13).

As seen above, although the complete removal of smoke might be out of reach, it seems that the technical means available in those days could remove a considerable portion of the smoke, or at least that many people at the time recognized that smoke could be abated. Thus, the Committee confirmed the existence of various technical approaches, the viability of the initial costs, the potential for smoke abatement, and especially, the active and positive adoption of technical means by factory owners. In other words, the Committee obtained the corroboration it needed to institute anti-smoke policies.

5.2.3   On the Effect of Saving Coal

However, the technical means of abating smoke had a further effect, "the saving of fuel". This effect was almost always mentioned in testimony on smoke-abatement means, and in a certain sense, greater importance was attached to it than to the smoke-abatement effect.

The fact that the technical means for smoke abatement could also save fuel was what the developers especially emphasized. When Gregson was asked by the Committee, "Are you aware of any practical means by which those nuisance can be removed?", he responded, "Yes", but pointed out that some of the older means increased fuel consumption. Conversely, he emphasized, his own furnace would reduce coal consumption by 10 % (Source 2, pp. 5–6). It appears that early on, the Committee concerned itself with costs and performance and did not have much in fuel savings. Needless to say, however, if the technical means for smoke abatement could be shown to save fuel, that would have a major significance in enforcing the adoption of smoke-abatement means by manufacturers. Thus, after Gregson's testimony, the Committee became interested in fuel savings and began to ask specifically for testimonies on this effect. Walker answered that his savings reached 25 %, Losh said that the use of his furnace "is connected with the saving of fuel, as well as the burning of fuel", and Moult, Parkes, and Wakefield testified that their furnaces saved about 25 % in coal. As already stated, since the technical means of the time were based on the principle of a perfect combustion of coal through increased air (oxygen) inflows to a furnace, it is to be expected that they would have implications for fuel economy as well. Brunton, who developed that expensive revolving grate, also placed strong emphasis on this economy. He said that his equipment promised 30 % or more in fuel economy, and insisted that the initial costs would be recovered through fuel economies in a little less than 18 months in London, or 3 years in Staffordshire (the difference being due to the gap in coal prices between London and Staffordshire) (Source 2, pp. 8–9, 12; Source 1, pp. 6, 12, 14–15).

Manufacturers testified on fuel economy. A distiller in London who was using Brunton's equipment said that he saved 38 % in fuel, and a sugar refiner in Liverpool who installed the same equipment in a furnace for a 32 hp steam engine testified that the amount of smoke became negligible, that his steam power increased significantly, that work which had conventionally required 1 h could be completed in 45 min, and that he had a savings of a quarter to a third in fuel. And he said that he was completely satisfied (Source 1, pp. 14–15).

5.3   Means for Preventing a Smoke Nuisance and Capitalists

As already stated, many of the manufacturers summoned to the public hearings used smoke-abatement means even before the inauguration of the Committee. Why was this?

One of the most significant motives was in fact fuel economy. The alum manufacturer referred to earlier adopted smoke-abatement technology in another of his factories, "by way of saving coals" first and foremost. And one sugar refiner said that he did not necessarily ask for Brunton's equipment for the purpose of smoke abatement at the beginning. Although he had introduced a large machine for sugar purification, he was faced with the problem that the machine could not perform at its full potential because of the shortage of heat in the furnace. He was therefore seeking methods for increasing heat, whereupon he learned of Brunton's equipment, which would not only increase the heat but "would also have the effect of burning the smoke" (Source 1, pp. 15–16; Source 2, p. 8).

Of course, fuel economy was not the manufacturers' only motivation for adopting smoke-abatement technology. They were also responding to frequent complaints from neighbors, or prosecutions under local Acts that made it possible to bring complaints against people who caused a smoke nuisance. Manchester and Birmingham had already enacted local laws on smoke regulation. A brewer in Salford was prosecuted under the Manchester Police Act, and in order to avoid being fined, he adopted Wakefield's furnace. Watt developed his technology at the request of Drinkwater in Manchester, who was a spinner, one of the oldest users of Watt's rotating engine, and was the target of smoke-complaints by his neighbors (Hills 1989, pp. 61–62). It seems, however, that these complaints and prosecutions had been the opportunity for these manufacturers to learn about the existence of smoke-abatement means, rather than their motive for using them. Once they knew about the possibilities, they chose to adopt the new means even where no complaints were in question, for the sake of the fuel economy they offered.

The above shows the following. (1) The technical means for smoke abatement had already existed to some extent and had been adopted by manufacturers before the Committee's inauguration. (2) The initial cost of installing apparatus for smoke-abatement was comparatively low. (3) Smoke could be reduced to some extent by using the available means. Simultaneously, (4) the use of smoke-abatement technologies saved fuel and enabled the eventual recovery of initial costs. (5) Although manufacturers had various reasons for introducing smoke-abatement apparatus, improved fuel economy proved an especially important motivation. Even capitalists who were not interested in smoke pollution were thus motivated to undertake of their own accord the improvements that also resulted in reduced smoke pollution. The pursuit of private profit (or fuel economy) on entrepreneurial principles could therefore be connected to the public benefit (of smoke prevention). Public benefit accorded with private profit.

6 Enactment of the Smoke Nuisance Abatement Act of 1821 and the Limit

6.1 Presentation of a Bill

The fact that some factory owners were positive about adopting technical means for smoke abatement, and that not only sufferers but capitalists, too, might benefit as a result, made Taylor submit the following report and decide to present an anti-smoke bill. The report says, "the Nuisance so universally and so justly complained of, may at least be considerably diminished, if not altogether removed" (Source 2, p. 3).

The bill says, "Whereas great inconvenience has arisen, and a great degree of injury has been and is now sustained by His Majesty's subjects, in various parts of the United Empire, from the improper construction as well as from the negligent use of Furnaces employed in the working of engines by Steam." Their recognition that smoke pollution had become a severe nuisance all over Britain is evident here, and smoke abatement is declared to be technically possible. "And whereas it has been ascertained, that the greater proportion if not the whole of the smoke issuing from such Furnaces, may, by a proper construction and due attention, be consumed, so as to prevent the Nuisance thereby arising." Furthermore, the bill claimed that some national measure was required to address the situation. That is, "And whereas by law, every such Nuisance, being of a public nature, is abateable [sic] as such by indictment; but the expense attending the prosecution thereof, have deterred parties suffering thereby, from seeking the remedy given by law" (Source 5, p. 1).

As already said, Britain had and has common law. And it was already possible to bring complaints of nuisance against smoke generators under common law. Nuisance was classified as either private or public. When a defendant was convicted of private nuisance, the plaintiff was granted the right to claim all damages arising from the defendant's acts, and also to claim an injunction for self-help. Furthermore, where a defendant continued to cause the nuisance deliberately or with malice, he was burdened with punitive damages. Defendants convicted for a public nuisance were condemned to a confinement of 2 years or less or a fine, at the judge's discretion, and the nuisance was prohibited by rule of the Attorney General. Moreover, a sufferer who suffered special damages beyond general inconvenience and displeasure was also able to claim damages. However, since nuisance trials required plenary rather than summary proceedings, their costs were high, and in cases of public nuisances, costs rose still higher because of the many witnesses and long time that was required (Katō 1979, pp. 271–272).

The bill sought to address these problems in its first clause, which made it lawful "for the Court before which any such indictment shall be tried, in addition to the judgment pronounced by the said Court in case of conviction, to award such Costs as shall be deemed proper and reasonable to the prosecutor or prosecutors, to be paid by the party or parties so convicted as aforesaid." The second clause of the bill adds that "if it shall appear to the Court before which any such indictment shall be tried, that the grievance may be remedied by altering the construction of the Furnace, or any other part of the premises of the party or parties so indicted, it shall be lawful to the Court, without the consent of the prosecutor, to make such Order touching the premises, as shall be by the said Court thought expedient for preventing the Nuisance in future, before passing final sentence upon the defendant of defendants so convicted" (Source 5, pp. 1–2).

6.2 Deliberation on the Bill: Resistance by the Mining and Iron Industry

The bill was presented to the House of Commons on April 18, 1821. At the beginning of the first reading, Taylor said, "Of the pernicious effect of such nuisances, no gentleman could be unaware. The steam engines productive of these nuisances were not only injurious to the health and comfort but even ruinous to the property of persons who happened to be resident in their vicinity." He then recounted the damage caused by smoke to himself and Jenkinson. He also introduced a clergyman who owned a school building that had been rendered unusable by smoke from a steam engine, and who could not afford to bring a lawsuit. Taylor noted that lawsuits were prohibitively expensive and that the most harmful of nuisances were often overlooked because the people living near the source of the smoke could not incur those costs. He concluded by saying that this situation would improve were the bill to pass (Source 4, 18 April 1821, col. 439–440).

At the first reading, Sir Matthew White Ridley, MP for Newcastle upon Tyne, was first to announce his full support of the bill, citing as an example the successful adoption of smoke-prevention measures in a Northumberland coal mine (Judd 1955, p. 318; Source 4, 18 April 1821, col. 441). Two other MPs, by contrast, though consenting to the meaning of the bill, demanded exemptions. John Hearle Tremayne, MP for Cornwall, and Matthew Wood, MP for the City of London, Alderman of the City of London, ex Lord Mayor of the City of London, and a malt-house owner, called for an exemption for the mining districts of Cornwall (Judd 1955, pp. 359, 381; Stenton 1976, p. 416; Source 4, 18 April 1821, col. 441). Taylor refused, stating that he was not considering any exemptions: "If the House should agree to introduce such a provision, I [Taylor] would abandon the bill altogether" (Source 4, 18 April 1821, col. 441). Since there was no objection to the meaning of the bill, it was sustained and sent on for a second reading.

In the second reading held on 7 May, four MP announced their full support of the bill. Charles Calvert, MP for Southwark and a brewer using steam engines, said that he had successfully adopted a smoke-prevention furnace. John Christian-Curwen, MP for Cumberland and the owner of steam engines, said that he had tried a new furnace, and that it had produced "ultimate saving" in fuel as well as reducing smoke. Joseph Marryat, MP for Sandwich, said that new furnaces were used successfully in London and elsewhere, and he supported the bill on the moral grounds that "No man had a right to annoy or poison his neighbors." John Smith said that the main sufferers were "poor people" and especially people in "the humble class" who got their livelihood by washing in London, and that if steam engines were established near their dwellings, they would be unable to carry on their occupation at all. He therefore insisted that the proposed bill was an absolute essential in order to safeguard their livelihoods (Judd 1955, pp. 140, 151, 271, 336; Stenton 1976, pp. 63, 353; Source 4, 7 May 1821, col. 536–538).

Nevertheless, claims for exemption were made also in the second reading. Staffordshire MP Edward John Littleton, who would become MP for South-Staffordshire in 1832 and who was probably, therefore, a spokesman for the iron industry, expressed the laissez-faire point of view that: "If the plan of consuming the smoke of Steam Engines was a good one, it would find its way without any legislative enactment; if a bad one it ought not to be forced upon the country", and he said, "In the south of Staffordshire there were above 2,000 steam engines; and in the neighboring countries at least 5,000 more. Parliament ought to hesitate before they imposed a compulsory expense and inconveniences on so many persons. If my [Littleton's] Hon. Friend should persist in the measure, I [Littleton] would propose that it should not extend to steam engines employed in smelting ores or minerals", and demanded exemption for the iron industry (Judd 1955, p. 258; Stenton 1976, p. 239; Source 4, 7 May 1821, col. 535–6). Thomas Wood, MP for Brecknockshire in South Wales from 1806 to 1847 (and probably also an iron-industry spokesman), likewise said, "representing as he [Thomas Wood] did a country where manufactures were carried on by steam-engines, he [Thomas Wood] felt it his [Thomas Wood's] duty to oppose proceeding any further with the bill in its present shape" (Judd 1955, p. 381; Stenton 1976, p. 416; Source 4, 7 May 1821, col. 537). Matthew Wood remarked that "Mr. Parkes's plan of consuming smoke was highly beneficial, but he [Matthew Wood] hope that the clauses of the bill would not extend to Cornwall, as it would there produce the most injurious effects", and again called for exemption for nonferrous metal mines in Cornwall (Source 4, 7 May 1821, col. 538).

As mentioned above, exemptions were demanded by spokesmen of the iron industry in South Staffordshire and South Wales, and of nonferrous metal mining in Cornwall. Both were among the largest industries using steam power, and had been consumers of steam power well before Watt's steam engine came to be used for machines in factories. The Cornwall mines, in particular, were long-time steam-power consumers, as is evident in the fact that Savery's fire-engine and Newcomen's atmospheric engine, patented in 1698 and 1705, respectively, had been developed for their use. The iron industry's use of steam power dated back to 1742, when Derby began using Newcomen's engine for pumping into water wheels. Even if

Littleton's numbers regarding South Staffordshire were exaggerated, there is no question that they were the most powerful steam users in the Industrial Revolution.

The iron and mining industries had no representatives in the Committee, and this appears to have driven them to make claims for exemption. The factory owners who were Committee members or witnesses came mainly from the textile or food industry. Thus, the Committee did not necessarily reflect the interests of all steam users in early nineteenth century Britain. It became clear several years later that the advantage to factory owners that derived from smoke-abatement measures, especially fuel economy, differed according to production conditions in the various industries.[5] To be sure, iron manufacture and mining had the right to demand exemptions from a bill that had been drafted without in any way taking their interests into account.

Not only were exemptions claimed at the second reading, but the bill's rejection, on account of technical difficulties, was demanded as well. Thomas Fowell, a brewer in London and MP for Weymouth, made a motion that the bill be referred to the whole committee 6 months later, saying that: "with an engine constituted as his [Buxton's] was, it was quite impossible to carry it into effect." Isaac Gascoyne, MP for Liverpool, seconded Buxton's motion, saying he had received a letter from a Liverpool manufacturer insisting that the new furnaces would lead to increases in both smoke and fuel consumption. Taylor, however, had no difficulty responding to Buxton's motion, probably because Buxton as in the brewing industry, regarding which many successful experiments had been presented in the Committee and before Parliament. He replied that, "If hon. Members would read the reports of the two committees who had considered the subject, they [Members] would be satisfied. He [Taylor] could produce the testimonials of persons who found the plan successful at little expense. The hon. gentleman (Mr. Buxton) must have employed a very clumsy engineer" (Judd 1955, pp. 138, 204; Stenton 1976, p. 60; Source 4, 7 May 1821, col. 536–538).

At the end of the second reading, Parliament put to the vote both Taylor's motion for passage of the second reading and Buxton's motion to reject the bill. The result was 83 votes for Taylor's motion and 29 for Buxton's (Source 4, 7 May 1821, col. 538). However, the final bill was revised, taking the opinions of Littleton and others into account, probably when it was brought before the whole committee for a debate on the details.[6] The third, and newly added clause, said, "the Provisions of this Act, as far as they related to the Payment of Costs and the Alternation of Furnaces, shall

---

[5]Iron manufacturers in particular differed from other manufacturers. They used the undesirable slack coal or 'screenings' from their own coal mines to fuel their steam engines. That coal was waste and worthless. They therefore could not expect to see profits from the fuel economies associated with smoke-abatement techniques. Other nonferrous-metal manufacturers closely related to coal mines were also in the same situation. This problem became clear in the investigations attending an anti-smoke bill in the 1840s (Source 6). For details, see Akatsu (2005).

[6]The reason to use 'probably' is that the record of deliberation over the bill in Whole Committee and the House of Lords does not exist in *Parliamentary Debates*. Ashby and Anderson assumed the revision of the bill in the House of Lords but did not show on what basis (Ashby and Anderson

not extend or be construed to extend to the Owners or Proprietors or Occupiers of any Furnaces of Steam Engines erected solely for Purpose of working Mines of different Descriptions, or employed solely in the smelting of Ores or Minerals, or in the manufacturing of the Produce of such Ores or Minerals on or immediately adjoining the Premises where they are raised" (Source 7, 1 and 2 George IV, c.41).

Although resistance by iron manufacturers and miners led to their exemption, the spokesmen of other industries were unexpectedly positive, as has been shown and explained above. The bill was sent to its third reading on May 10, 1821, and enacted.

6.3 Effects and the Limits of the Act

How much effect did this Act, drafted by a landlord and enacted with the support and cooperation of certain capitalists, have in the prevention of smoke nuisances? To be sure, the Act's effects were limited, as is often the case with pioneering legislation. For example, John Edward Nassaw Molesworth, a witness on the "Select Committee on Smoke Prevention", testified in 1843 that "the Act of Parliament commonly called M. Angelo Taylor's Act is inoperative" (Source 3, p. 77). But Parkes, who was summoned to the same committee, said: "Mr. Michael Angelo Taylor's bill was going on and it frightened the manufacturers; and for a while it frightened them into the adoption of my plan; the pressure from without (if I may so say) produced some good to me and to them too; but as soon as the pressure slackened, the masters did not care how much they smoked their neighbours or anybody else" (Source 3, p. 165). That is, the Act had the effect of preventing smoke nuisances for some time after being enforced. The problem was that its effect did not last long.

There were several reasons for this. First, the Act only enabled lawsuits for smoke nuisances under common law, while retaining a laissez-faire character in the relationship between the government and the economy or market. That is, although the purpose of the Act was for the government to compel manufacturers to adopt smoke-abatement means and in that sense to interfere in private enterprise, the government's role in enforcing the Act could hardly be said to have been active. The reason for this is that lawsuits against nuisances under common law are brought on the initiative of individual complainants, and they were supposed to take the initiative in demanding compensation or relief. If a sufferer did not challenge the smoke-maker, the smoke-makers would not be accused of creating a nuisance. The Act was a form of laissez-faire legislation, compared with the strict and interventionist acts that followed, which permitted the government to prohibit

---

1981, p. 5). However, I think there is a greater likelihood that the bill was revised in the course of detailed discussions in Whole Committee.

smoke nuisances and to prosecute on evidence given by appointed inspectors, regardless of whether any sufferers stepped forward.[7]

Why, then, did Taylor choose to adopt legislation that was non- interventionist, even though he had sufficient corroboration for regulation in the form of some support from entrepreneurs and the private and public benefit that would result from smoke-abatement? The reason becomes clear when we look at the kind of people Taylor was trying to assist through the Act. The sufferers whom he took up in the deliberation process of his bill were himself, Lord Liverpool, a clergyman, a gentleman, and property-holders who suffered from smoke. Taylor assumed that the principle sufferers were citizens who held certain properties. With support form the Act, they could bring appeals against smoke-makers without depending on governmental authority, and could win relief. They were probably the ones described in Parkes' testimony as those who had applied pressure to manufacturers in the early days of the Act's enforcement.

However, changes in social conditions had emasculated the Act by the 1840s, much as Molesworth said. Lawsuits by rich citizens decreased, and pressure from them diminished, because they had begun to escape from city-centers to the suburbs. For rich citizens, it was possible to get away.[8] Meanwhile, laborers, "poor people", and people in "the humble class", as Smith of London described them, came to be concentrated in areas near factories that had steam engines, and it was hard for them to obtain relief through the Act. As was pointed out by witnesses Henry Thomas De la Beche and Lyon Playfair, who investigated smoke-nuisances nationwide at the request of the government in 1846: "Your Lordship will remember that smoke is a public nuisance under the common law, and, as such, may be proceeded against by indictment. But the procuring evidence on a public nuisance, sufficient to insure conviction, is so great that the common law of nuisance is inapplicable to prevent this increasing evil. The persons immediately subject to the nuisance of a smoky factory, for example, are in many cases dependent upon that factory for employment..." (Source 8, p. 4). It is easy to understand that laborers would have difficulty bringing appeals against factory owners. The pressure on smoke-makers weakened, and the Act lost its power. Since Taylor had relied on the common-law

---

[7]Nuisances, including smoke nuisances, on statute, for example, a smoke nuisance under the Public Health Act of 1936, are mainly prosecuted by public institutions (local authorities). For details, refer to (Katō 1979, pp. 308–311). However, in the case of a nuisance under common law, the right of action is mainly given to the sufferers themselves. For details, refer to (Katō 1979, p. 289; Mochizuki 1990, p. 221). But, in the case of a common-law nuisance, when it is a public nuisance, the Attorney General (a cabinet member) is in charge of the prosecution and therefore it is the government that is intervening. Still, the role of sufferers themselves as witnesses is critical to obtaining conviction in the trials of public nuisances. This is clear from the testimony of the government-appointed inspector referred to in section 5.

[8]It was pointed out frequently in the 1840s that citizens had escaped to the suburbs. For example, Darnton Lupton, Mayor of Leeds, said, "every one who can, is going out of town", and he himself had escaped to suburbs three miles outside the city of Leeds, because of the annoyance of smoke-damage to household goods (Source 6, pp. 33–34).

approach of responding to those individuals who came forward with complaints, nuisances were inevitably neglected when sufferers declined to do so.

Air pollution became a nationwide public nuisance during the Industrial Revolution. Rich citizens were among the sufferers from smoke and constituted a part of the many and unspecified people who needed relief. But the remainder of the people, namely the laborers in the new industrial cities and the lower classes with whom Taylor did not concern himself, were not given the means of relief. The time had not yet come for demands that the government prosecute smoke-makers, rather than relying on individuals to bring complaints on their own behalf.

Despite these limits, however, the enactment of the Act was significant. First, the Act did have meaningful effect, at least in the early days of its enforcement. Second and more important, the Act led to the legislation later enacted for similar reasons. It was through the enactment of this Act that Parliament gave public recognition to air pollution as an important social problem. This Act represented a public declaration that the issue would henceforth be addressed through the enactment of Statutes and deliberations on Bills, and public confirmation was made of the fact that some capitalists favored smoke-abatement, and that smoke-abatement also offered potential fuel economies. In fact, after the enactment of the Act, air pollution came under frequent and regular consideration as one of the most important subjects for discussion by Parliament.

7 Conclusion

This article has shown the following:

1. During the Industrial Revolution, air pollution caused by coal smoke became a social problem (or public nuisance) nationwide, and was recognized as such.
2. Earlier studies maintained that it was a humanitarian politician who first proposed measures for improvement, but it was in fact a landlord with a direct interest in the matter who did so.
3. Although previous studies have uniformly regarded factory owners only as polluters, some factory owners (especially those in the textile and food industries) were themselves aware of the economic damage caused by smoke and took an active and constructive part in developing approaches to it.
4. The technical means used for smoke-abatement in those days not only could diminish the amount of smoke, but could also decrease the consumption of fuel. It was therefore possible to advance the battle against air pollution without greatly injuring the profitability of factory owners who were asked to adopt these methods.
5. The Smoke Nuisance Abatement Act of 1821 was enacted against this background, and with the positive support of landlords and some factory owners. However, in the process of deliberations on the Bill, certain industries were exempted from the Act in response to their concern about their own interests.

The reason the legal measures to control pollution became vulnerable was not because of bigoted resistance by a majority of factory owners or because of some technological barrier, but because of a conflict of interests among industries, the details of which were not apparent at the time.

6. Due probably to the limited concern of the person who brought the bill, the Act aimed only at the promotion of lawsuits under common law, limiting the main candidates for relief to rich citizens who could bring accusations against factory owners. As a result, even though the Act was effective early in its enforcement, it lost its effect as rich citizens began to move out to green suburbs and the areas near factories were populated by labourers who were dependent for their livelihoods upon those very factories.

7. Nevertheless, the Act's passage demonstrated that this problem was the responsibility of Parliament, and led to the subsequent enactment of other related Acts.

When in the 1840s the Act's weakness began to be noted, Parliament was prompt to take corrective action on the basis that legal measures were already indicated. A humanitarian politician took on the cause of smoke-abatement. He took action to relieve poor workers from the damage caused by smoke, without reference to his own interest. The measures he demanded were regulations to ensure that inspectors appointed by local authorities would prosecute smoke-makers regardless of the existence or number of sufferers. This was much more severe than Taylor's laissez-faire approach. Playing the lead role was the owner of a large cotton manufactory in Manchester. He was the one substantially to draft what was a strict bill prohibiting smoke. The demands for smoke-abatement measures by factory owners that were already present in the 1820s became still stronger and clearer in the 1840s.

However, their efforts did not succeed, and it would be somewhat longer before similar regulations were put into practice. Among the causes of the setback were the same conflicts of interest among industries that had not been clear in 1820. For legislation to regulate pollution, differences and conflicts of interest among industries is a critical problem comparable to the confrontation between enterprises and sufferers. This is a subject for further consideration that has not been addressed sufficiently in previous studies.

Parliamentary Papers

Source 1: Select Committee on Steam Engines and Furnaces, Parliamentary Papers, II, 1820.
Source 2: Select Committee on Steam Engines and Furnaces, Parliamentary Papers, VIII, 1819.
Source 3: Select Committee on Smoke Prevention, Parliamentary Papers, VII, 1843.
Source 4: Parliamentary Debates.
Source 5: Bill to give greater facility in prosecution and abatement of Nuisances arising from Furnaces used in working of Steam Engines, Parliamentary Papers, II, 1821.
Source 6: Select Committee on Smoke Prevention, Parliamentary Papers, XIII, 1845.
Source 7: The Statutes of the United Kingdom of Great Britain and Ireland.
Source 8: Report on Means of obviating Evils arising from Smoke by Factories and Works in Large Towns, P. P., XLIII, 1846.

References

Akatsu, Masahiko. 2005. Jūkyūseiki-chūyō no Igirisu ni okeru taikiosen mondai: 1844nen engaikinshihō wo megutte (The air pollution problem in mid-nineteenth century Britain: The Smoke Prohibition Bill of 1844). *Rekishi to Keizai* (The Journal of Political Economy and Economic History) 47(4):17–32.

Ashby, Eric, and Mary Anderson. 1981. *The politics of clean air*. Oxford: Oxford University Press.

Brimblecombe, P. 1988. *The big smoke: A history of air pollution in London since medieval times*. London: Routledge.

Cannon, J. 1997. *The Oxford companion to British history*. Oxford: Oxford University Press.

Flick, Carlos. 1980. The movement for smoke abatement in 19th-century Britain. *Technology and Culture* 21(1):29–50.

Gimpell, J. 1975. *La revolution industrielle du moyen age*. Paris: Editions du Seuil.

Hills, R.L. 1989. *Power from steam: A history of the stationary steam engine*. Cambridge: Cambridge University Press.

Horibe, M. 1975. Igirisu Manchesutā no taikiosen taisaku (Anti air pollution policies in Manchester). *Kōgai Kenkyū* (Research on Pollution) 5(1):60–68.

Judd, G.P. 1955. *Members of parliament 1734–1832*. New Haven: Yale University Press.

Kanefsky, J., and J. Robey. 1980. Steam engines in 18th-century Britain: A quantitative assessment. *Technology and Culture* 21(2):161–186.

Katō, Ichirō (ed.). 1979. *Gaikoku no kōgaihō* (Anti-pollution laws in foreign countries), vol. 1. Tokyo: Iwanami Shoten.

Kurachi, Mamoru. 1982. *Kankyō kagaku gairon* (Introduction to environmental studies). Tokyo: Kyōritsu Shuppan.

Landes, D.S. 1977. *The unbound Prometheus: Technological change and industrial development in Western Europe from 1750 to the present*. Cambridge: Cambridge University Press.

Mitchell, B.R. 1988. *British historical statistics*. Cambridge: Cambridge University Press.

Mochizuki, Reijirō. 1990. *Eibeihō* (Anglo-American Law). Tokyo: Seirin Shoin.

More, C. 1997. *The industrial age: Economy and society in Britain 1750–1995*. London: Longman.

Muraoka, Kenji. 1980. *Vikutoria Jidai no seiji to shakai* (Politics and society in Victorian Britain). Kyoto: Minerva Shobō.

Musson, A.E. 1976. Industrial motive power in the United Kingdom, 1800–70. *Economic History Review* 29(3):415–439.

Musson, A.E. 1978. *The growth of British industry*. London: B. T. Batsford.

Nakamura, Hidekatsu. 1961. *Igirisu Gikai seiji no hattatsu* (Development of British Parliament). Tokyo: Shibundō.

Nakamura, Hidekatsu. 1976. *Igirisu Gikai Seiji-shi ronshū* (Collected papers on the history of British Parliament). Tokyo: Tōkyō Shoseki.

Nef, J.U. 1966. *The rise of the British coal industry*, vol. 1. London: Cass.

Ōba, Hideki. 1979. *Kankyō mondai to sekaishi* (Environmental problems and world history). Tokyo: Kōgai Taisaku Gijutsu Dōyūkai (Association for Anti-pollution Technologies).

Okada, Tomoyoshi. 1987. *Keizaiteki jiyūshugi* (Economic liberalism). Tokyo: Tokyo Daigaku Shuppankai.

Sekiguchi, Yoshiyuki, and Jun'ichi Umetsu. 1995. *Ōbei keizaishi* (Economic history of Europe and the United States), 3rd ed. Tokyo: Hōsō Daigaku Kyōiku Shinkōkai (Foundation for the Promotion of the Open University of Japan).

Stenton, M. 1976. *Who's who of British Members of Parliament, vol. 1, 1832–1885*. Hemel Hempstead: Harvester Press.

Sugai, T. 1974. Eikoku ni okeru taikiosen no rekishi (History of the air pollution problem in Britain). *Kōgai to Taisaku* (Journal of Environmental Pollution Control) 10(10):87–90.

Sweet, R. 1999. *The English town, 1680–1840: Government, society and culture*. London: Longman.

Takei, Yoshiaki. 1984. *Igirisu no chiiki to shakai* (Region and society in Britain). Tokyo: Ochanomizu Shobō.

Tamura, Kōichi. 1965. Taikiosen bōshi kankeihō (On acts for air pollution control). *Jurisuto* (Jurist), 326:64–68.

Te Brake, W. H. 1975. Air pollution and fuel crises in preindustrial London, 1250–1650. *Technology and Culture* 16(3):337–359.

Webb, Sidney, and Beatrice Webb. 1963. Statutory authorities for special purposes: With a summary of the development of local government structure. *English Local Government*, vol. 4. London: Cass.

Part II
Book Reviews

Chapter 5
Review of Seiichi Andō, *Kinsei Kōgaishi no Kenkyū* (The History of Environmental Pollution in Early Modern Japan)

Yoshikawa Kōbunkan, Tokyo, 1992

Yuko Okada

This is an ambitious work of empirical study by Seiichi Andō, who has been a major contributor for many years to research on the socio-economic history of early modern Japan.

Historical research on environmental pollution in Japan began, as the author points out, during the 1960s years of high economic growth, and developed in earnest in the 1970s, reflecting the surge of public opinion against environmental pollution. While valuable research was conducted in that period, exposing information that had till then been hidden, the principal purpose of these studies was the analysis of the relationship between capitalism and pollution problems. Research continued to progress in the 1980s and 1990s, when public concern over the "environment" shifted from "the pollution problem" to "ecology", but the focus remained mostly on the late modern period: only a few case studies were made of the early modern period. It was in this context that this work was published, becoming the first comprehensive study of environmental pollution in early modern Japan. The publication was a significant event and the author deserves our deep respect.

Only a short summary of the content and a few comments will be presented here due to the reviewer's limited knowledge of the early modern period.

The author argues in his preface and introduction that, given the global scale of today's environmental destruction, "it is natural that the history of hitherto neglected 'negative' or 'adverse legacies' will become more and more important and that

This is a translation of a book review that originally appeared in *Shakai Keizai Shigaku* 58(4) (November 1992), pp. 114–116.

Y. Okada (✉)
Faculty of Economics, Kyūshū Kyōritsu University, Kita Kyūshū, 807-8585, Japan
e-mail: y-okada@kyukyo-u.ac.jp

© Socio-Economic History Society, Japan 2015
S. Sugiyama (ed.), *Economic History of Energy and Environment*, Monograph Series of the Socio-Economic History Society, Japan, DOI 10.1007/978-4-431-55507-0_5

environmental history, or the history of environmental pollution, will attract more attention" (Introduction, p. 2). Casting doubt on the widely accepted view that the history of Japan's environmental pollution began with the Ashio Copper Mine Mineral Pollution Incident, Andō criticizes researchers of environmental pollution as well as historians for having "either denied the existence of environmental pollution in early modern Japan or neglected it ... even though at least some studies on early modern pollution have already been produced" (Introduction, p. 5). In order to demonstrate that "environmental problems in the early modern period were not just isolated incidents", he presents "historical evidence of environmental pollution in every district of the country" (Preface, p. 2 and Introduction, p. 5). According to the author, environmental pollution can be traced through historical evidence back to the sixteenth century, and still further even to the mid-eighth century. He defines environmental pollution as "environmental destruction, caused mainly by human economic activity, that damages the local community, either directly or indirectly", and regards it as "a phenomenon that is not particular to a given economic system, even if in its details it takes on the particular attributes of that system" (Introduction, p. 2).

Chapters 1, 2, 3, 4, and 5 cite environmental problems district-by-district. Chapters 6 and 7 take up specific problems caused by water-wheel use and coal mining. Chapter 7 focuses mainly on northern Kyūshū and is in that sense a case study of environmental pollution in the Kyūshū district. The author took pains to collect historical materials covering almost the entire country from Tōhoku to Kyūshū except for the Tōkai district, but more than 60 % of the volume addresses western Japan, as the author himself admits that "his research focused on his own region" (Preface, p. 2). Among other points the reviewer finds somewhat unsatisfactory about this book is the paucity of figures and tables. Explanations with figures and tables would have helped readers understand the positional relationship of the various incidents and also the gist of the author's arguments.

Although this review does not discuss individual cases in detail, it should be noted that, in addition to the pollution caused by mining, which affected areas well beyond domainal (*han*) borders, urban river pollution, the development of new rice fields and saltpans, and the use of water wheels were also forms of environmental pollution that affected local communities. As the author emphasizes, even the development of new rice fields had its negative effects. It is also an important discovery that water wheels generated environmental problems. In addition, the author shows that the protest movements carried out by local residents against environmental pollution between the seventeenth and nineteenth centuries were based on precedents and information they gathered about other areas suffering the same problems. This is a particularly important finding on information dissemination, given that the flow of this information coincided with the spread of agricultural tools and techniques from the Kinki district to the rest of the country. Other interesting findings include the fact that in the seventeenth and eighteenth centuries, mining and water-wheel operations were restricted to certain times or even totally banned on many occasions.

Chapter 8, "Debates on Environmental Pollution in the Early Modern Period", analyzes how intellectuals at the time observed and debated environmental pollution. The author maintains that "in the early modern period it was considered highly important to maintain nature's balance, and environmental pollution was taken much more seriously than we assume it was" (p. 379). The final chapter states that, "environmental problems in the early modern period fell in the category either of 'development pollution' or of 'industrial pollution'" (p. 384). Addressing the arguments in the previous eight chapters under the headings "Origins of Environmental Pollution", "Types of Environmental Pollution", "Measures against Environmental Pollution", "Compensation for Environmental Pollution", and "Peasant Revolts against Environmental Pollution", the author stresses that just as environmental pollution can be traced back to the sixteenth century, "countermeasures, compensation, and peasant protests can also be traced back to the seventeenth century", and that "the number of such cases increased rapidly in the nineteenth century, reflecting the development of industrialization in early modern Japan" (p. 397).

The above is the gist of this study. This work deserves praise for amassing a rich array of historical evidence to demonstrate that "environmental problems were not isolated incidents in early modern Japan but rather were phenomena common to every district of the country" (p. 383). This work is the first comprehensive research on the history of environmental pollution in early modern Japan, and it is a *tour de force*. Furthermore, it is valuable as a collection of historical materials, and will without question become the foundation for future investigations in this area.

A few problematic points require mention, however. First, the term "ecosystem", which the author uses in defining the concept of "environmental pollution", is not sufficiently clear. On page 379, for example, the author writes that, "The natural balance deteriorated . . ." However, the reviewer cannot find a corresponding case in the citations of district-by-district environmental problems in Chaps. 1, 2, 3, 4, 5 and 7.

Next, although the author discusses environmental pollution in relation to industrialization (pp. 25, 209, 236 and 384), his arguments are not sufficiently persuasive, because he does not clearly present the actual conditions of industrialization. Policies taken by the central and local governments (*bakufu* and *han*) are mentioned in Chap. 8. However, a more penetrating analysis would have been possible had the pollution problems been discussed in relation to such factors as the structural changes taking place in the Tokugawa system, the increases in population, cultivated area and land productivity, and the particular characteristics of local economies. The author's view on these issues would also have been instructive, given their connection to fundamental questions about the relationship between environmental pollution and economic development during the transition from the early modern to the modern period and to how that relationship affected measures to "solve" environmental problems.

The author set himself the task of demonstrating that "environmental pollution . . . was a common phenomenon in early modern Japan" (Postscript, p. 399), and he succeeded admirably. There is no doubt that this work has established a foundation for historical investigation into early modern Japan's environmental pollution. It will be our task as well as the author's to carry forward with both theoretical and empirical investigations in this field.

Chapter 6
Review of Takeo Kikkawa, *Nihon Denryokugyō Hatten no Dainamizumu* (Dynamism in the Development of Japan's Electric Power Industry)

Nagoya Daigaku Shuppankai, Nagoya, 2004

Ryoshin Minami

This is a comprehensive study of the history of Japan's electric power industry. It achieves the status of a magnum opus not only because it comes to more than 600 pages, but also because it represents research of the highest quality. The author is a leading expert in the field, who has extensive background in writing and editing histories of electric-power companies. This book is expected to attract a great deal of attention from both economic and business historians, as is another monumental work by the same author (Kikkawa 1995).

1 Changes in the "Autonomy" of the Industry

This study is most remarkable for the fact that its research aims and analytical framework are clear and that its entire argument is therefore extremely lucid and persuasive.

The study aims to examine the dynamism of Japan's electric power industry by analyzing changes in the "autonomy" of the industry. "Autonomy" is here defined as "the complementarity achieved by being both a private company and a public utility", or in other words, the ability of the utility "to effect the secure supply of electricity at a low price through rationalization measures undertaken by privately-owned and privately-run companies" (Introduction).

This is a translation of a book review that originally appeared in *Shakai Keizai Shigaku* 71(1) (May 2005), pp. 84–86

R. Minami (✉)
Hitotsubashi University, Tokyo, Japan
e-mail: ryominami@nifty.com

As a private enterprise, an electric power company has profitability as its main aim. As a public utility, however, it has an obligation to society and the economy. This is why the study's analytical focus is on the question of "autonomy". As to how the "autonomy" of the industry changed in the course of development, the author's analysis can be summed up as follows:

The early years of the industry (1883–1906) were characterized by fierce competition among numerous private companies. This competition did not, however, meet the public interest by supplying cheap electricity, because of the cost effect of soaring coal prices on the thermal-power generation that produced most of the electricity. In other words, although the industry developed dynamically in those years, it did not demonstrate "autonomy" (Chap. 1).

In the next phase of the industry's development (1907–1931), electricity prices did come down to a certain extent, due to a combination of severe ongoing competition, referred to as "the electric-power battle", and the concurrent development of hydroelectric power generation. As the financial condition of electricity retailers deteriorated, however, the electricity supply system became unstable. The "supply of cheap electricity" did not prove compatible with the establishment of a "stable supply of electricity" (Chap. 2).

The adverse effects of the "electric-power battle" led power suppliers to adopt measures for self-regulation and to form the League of Electric Power Companies (a cartel organization) in 1932. Electricity prices were set low relative to other prices and electric power company operations stabilized. In other words, this period saw the early development of "autonomy" (Chap. 3).

The state took control of the electric power industry in 1939, and power companies accordingly lost their status as private enterprises. "Autonomy" was out of question and the industry lost its dynamism (Chap. 4).

The electric power industry was reorganized in 1951 and nine private companies were granted regional monopolies. Thus established was the system that has persisted until today. Between 1951 and 1973, roughly the period of Japan's high economic growth, demand for electricity increased rapidly and electric power companies enjoyed ongoing stability. Despite the presence of regional monopolies, competitive rivalry persisted among the power companies, compelling them to shift from hydroelectric to thermal power generation, which used oil more intensively than coal, and leading eventually to a reduction in the cost of electricity. This proved a golden period for the industry, in which "autonomy" functioned most effectively within a context of high levels of dynamism (Chap. 5).

The oil crisis of the early 1970s, however, forced the electric power companies to shift their source of energy from oil to natural gas and nuclear power. The nine companies raised electricity prices in unison. Cheap electricity became a thing of the past. The electric utilities instituted simultaneous price hikes again and again and took united action against the anti-nuclear movement; competition among them declined, and they formed ever closer ties with the government. As they "lost their character as private companies", the industry itself lost much of its dynamism (Chap. 6).

Privatization accelerated and public pressure intensified to close the gap between prices at home and abroad, prompting the revision of the Electricity Business Act in 1995 and 1999, which marked the beginning of the electricity deregulation. Electric power companies responded by reducing electricity prices in unison. In this sense, the "supply of cheap electricity" was achieved to a certain extent, and deregulation can accordingly be regarded as having spurred the revival of the industry's dynamism and "autonomy" (Chap. 7).

2   The Meanings of "Autonomy" and "Dynamism"

As has been shown above, this book astutely analyzes the 120-year history of the electric power industry through the framework of "autonomy". This approach will be valuable as a guide to a new direction in business history.

The word, "autonomy", however, needs to be used with care. Akifumi Nakase has raised a question in this regard, arguing that although the original meaning of "autonomy" is "to regulate one's own behavior by self-control", Kikkawa gives a different meaning to the word (Nakase 2002, p. 77). This is a legitimate point. The confusion could have been avoided had the expression, "the compatibility of being both a private company and a public utility", been used instead of "autonomy".[1]

The use of the word "dynamism", which is contained in the very title of this book, also requires special attention. Although it is hard to define this word rigorously, it is usually understood to mean "to develop rapidly, responding swiftly and boldly to change." (China's economy today is an example of dynamism as the national level.) If "dynamism" is defined this way, it can be measured by the industry's growth rate, which would be expected to have been high in the period of "dynamism" between 1951 and 1973. The point, however, is how that "dynamism" is generated and under what circumstances it recedes. Moreover, how is "dynamism" related to "autonomy" (regardless of whether the term is used in the author's sense or in its original sense)? Analyzing the history of a specific industry from this point of view, whether the electric power industry or any other, may prove a breakthrough in the research of industrial history.

---

[1] If 'autonomy' and its antonym, 'heteronomy' are defined respectively according to their original meanings, as change within an industry caused by 'internal factors' and that caused by 'external factors', 'heteronomy' appears to be a better description of the electric power industry case. For example, electricity deregulation did not happen at the behest of the industry, but rather as part of the Nakasone government's policy of deregulation and a free market economy. Moreover, there is no denying that the policy was influenced by Thatcherism in Britain and Reaganomics in the United States.

3 The Significance of the Reorganization of the Electric Power Industry in Japan's Economic History

Another characteristic of this work is the richly empirical and minutely detailed analysis of activities of those who contributed to the development of the industry, from which there is a great deal to be learned.

The section concerning Yasuzaemon Matsunaga especially stands out. It explains that in his 1928 *Personal Views on the Control of Electric Power*, Matsunaga envisioned the reorganization of the Japanese electric power industry, a vision that was to bear fruit in the 1951 reorganization of the industry. This view holds that the period of state control over the electric power industry between 1939 was a "long detour" (p. 11). This is an intriguing view in that it claims a certain continuity between the periods before 1938 and after 1951, which makes it part of the major debate among Japanese economic historians as to whether there was continuity or discontinuity between prewar and postwar Japan (Minami 1995). In my view, the industry's reorganization was part of the GHQ's policy to democratize Japan's socio-economic system. It is true that the GHQ adopted Matsunaga's ideas about the regional monopolies, but it was owing to the power of the GHQ that state control was terminated and the industry reorganized (A similar argument can be seen in Hashimoto 1996, p. 110). In this sense, the reorganization of the industry is an example of discontinuity.

4 Was the Electric Power Industry Japan's Leading Industry?

The electric power industry developed smoothly from the end of the Russo-Japanese War in 1905 until the late 1920s. The author adopts Takafusa Nakamura's argument in stating that "The electric power industry functioned as Japan's leading industry" (p. 78). In the 1920s, electric power industry contributed to national economic growth (the industry's share of the real NDP increase) at a rate that reached as high as an average of 17.3 %.[2] This does not necessarily mean, however, that the industry can be characterized as the economy's "leading" industry or engine. The electric power industry's contribution rate was outsize in part because of the stagnation many industries in the manufacturing sector experienced during the 1920s on account of the economic recession. The electric power industry's contribution rate is much lower in the other periods in question: 0.2 % for the 1890s, 1.1 % for the 1900s, 2.5 % for the 1910s and only 1.2 % for the 1930s.

---

[2]The calculation of the NDP is based on the prices of 1934–1936 (Ōhkawa et al. 1974, pp. 226, 230).

The electric power industry provided the industrial world with cheap electricity and contributed significantly to Japan's industrialization, enabling the shift from steam to electric power at large companies and the shift from human to machine power –in other words, mechanization– in small and medium-sized businesses (Minami 1976). It would be more accurate, in my view, to describe the industry's role not as "leading" economic development, so much as "supporting" it as a form of "social infrastructure".

References

Hashimoto, Jurō. 1996. Book review on Kikkawa Takeo, Nihon denryokugyō no hatten to Matsunaga Yasuzaemon. *Keizaigaku Ronshū* (Journal of Economics) 62(3):107–110.

Kikkawa, Takeo. 1995. *Nihon denryokugyō no hatten to Matsunaga Yasuzaemon* (Yasuzaemon Matsunaga and the development of Japan's electric power industry). Nagoya: Nagoya Daigaku Shuppankai.

Minami, Ryōshin. 1976. *Dōryoku kakumei to gijutsu shinpo* (The power revolution and technological developments). Tokyo: Tōyō Keizai Shinpōsha.

Minami, Ryōshin. 1995. Nihon keizai no sengo fukkō ga shisasuru mono (What does Japan's postwar economic recovery suggest?). *Keizai Seminā* (The Economics Seminar) 488:30–33.

Nakase, Akifumi. 2002. Nihon no denryoku kaisha no kyōkyū-sekinin tassei to karyoku kaihatsu (The fulfillment of Japanese electric power companies' responsibility for electricity supply and the development of thermal power generation). *Keiei Kenkyū* (Management Studies) 52(4): 77–100.

Ōhkawa, Kazushi, et al. 1974. *Kokumin shotoku* (National income). Tokyo: Tōyō Keizai Shinpōsha.

Chapter 7
Review of Yoshihiro Ogino (ed.) *Kindai Nihon no Enerugī to Kigyō Katsudō: Hokubu Kyūshū Chiiki o Chūshin to Shite* (Energy and Corporate Activities in Modern Japan: The Case of Northern Kyūshū)

Nihon Keizai Hyōronsha, Tokyo, 2010

Naoki Tanaka

Professor Yoshihiro Ogino was one of the speakers at the 46th Annual Conference of the Socio-Economic History Society held at Kyūshū University in May, 1977, which explored the theme of "Energy and Economic Development". Thirty years later, he has once again tackled the problem of "energy and economic development" in three key areas: "energy", "corporate activities" and "regional economy".

Each article in this collection is based on a case study conducted in a specific region, mainly in northern Kyūshū. The collection is edited with an intention to "highlight, in relation to energy and corporate activities, economic and industrial problems for the Kyūshū region and characteristics of the multilayered industrial structure in each historical period" (p. v). Three aspects are shared among the researchers in this project. The first is the structural relation between the "central" and the "regional", called the "structuralization" of the central and the regional. Secondly, the Japanese economy is analyzed from the perspective of "regional development". Thirdly, "corporate activities" of a regional economy are analyzed comprehensively (p. ii). This book is divided into two parts. Part One "Energy" (Chaps. 1, 2, 3 and 4) includes studies on coal and electricity. Part Two "Corporate Activities" (Chaps. 5, 6, 7, 8 and 9) includes studies on a railway company, a regional bank, the management of an engineering works company and a trade

This is a translation of a book review that originally appeared in *Shakai Keizai Shigaku* 77(3) (Novemeber 2011), pp. 135–137.

N. Tanaka (✉)
Nihon University, Tokyo, Japan
e-mail: tanakan1023@gmail.com

© Socio-Economic History Society, Japan 2015
S. Sugiyama (ed.), *Economic History of Energy and Environment*, Monograph Series of the Socio-Economic History Society, Japan, DOI 10.1007/978-4-431-55507-0_7

association. These studies cover a long time span stretching from the early Meiji period (ca.1868–1872) to World War II. The content of each chapter can be summed up as follows.

Chapter 1, "The Sales Tendency of Miike Coal in the Interwar Period", by Mitsuru Kitazawa is an empirical study on the sales of Miike coal. Based on the fact that Miike coal had a very different quality from other domestic coal, this study aims at "drawing a concrete picture of the markets of Miike coal". Miike coal's sales strategy was altered during the interwar period. Whereas previously Miike coal had been mostly transported to domestic markets or exported, it came to be used as bunker coal, exported or consumed locally. This shift was made possible for the following reasons: export was outside the cartel regulation; the Mitsui Mining Co. and the Mitsui Bussan adopted the strategy to use Miike coal more intensively as fuel for foreign ships and coal mines; the technological development made it possible to use Miike's coal dust as fuel for domestic ships; local coal consumption increased rapidly because of the development of the coal chemistry industry in the Ōmuta area.

Chapter 2, "The Management of the Silk Reeling Business and the Fuel Problem", by Kazue Enoki examines the fuel problem of the silk reeling industry in the prewar period. The consumption of coal has been a relatively neglected subject of research. Based on the understanding that the fuel cost constituted an important part of the total production cost in the silk reeling business, the following two points are investigated. First, the trend of the fuel consumption in the silk reeling industry is examined. During the 1930s the volume of coal consumed to produce a unit of silk was reduced by half, although the total volume of silk produced remained almost unchanged. Secondly, a case study of the Gunze Silk Reeling Co. is carried out, in order to examine why the above-mentioned change occurred. Many large silk reeling companies started to take measures to reduce the fuel costs during the 1920s. Consequently, small silk reeling businesses were largely eliminated. Thus, in part due to fuel-related problems, the silk reeling industry was reorganized.

Chapter 3, "The Regional Bias in the Spread of Coal Kilns: the case of the Arita pottery Industry', by Hidetoshi Miyaji studies the introduction of coal kilns into major pottery-producing areas: namely, the Tōnō area in Gifu prefecture, the Seto area in Aichi prefecture and the Arita area in Saga prefecture. Coal kilns were actively introduced to the Tōnō and Seto areas from the 1910s onwards. As a result, the fuel cost was reduced and these areas consolidated their status as major pottery-producing areas. In the Arita area, however, attempts to introduce coal kilns proved to be futile on account of the nature of the coal produced in northern Kyūshū. Consequently, the Arita pottery industry was forced to specialize in the production of high-quality goods, and thus Arita's status as a pottery-producing area declined.

Chapter 4, "The Supply and Demand of Electricity in the Ōmuta Area in the Prewar Shōwa Period: the Case of the Electricity Strategy of the Mitsui Mining Miike Station", by Yoshihiro Ogino examines in detail the supply-demand balance of electricity in the Ōmuta area of Fukuoka prefecture where the Mitsui Mining Miike Station was located. In order to supply the electricity which the company needed in this area, Mitsui Mining initially decided that it would depend chiefly on in-house power generation and that power purchase would be a supplementary energy source. It turned out, however, that the company came to consume much more electricity than had been expected. Therefore, it had to find a new way to cope with its rapidly increasing demand for electricity. Eventually, in January 1935, Mitsui Mining was compelled, by the administrative guidance of the Kumamoto Communications Bureau, in the Ministry of Communications, to found the Kyūshū Kyōdō Thermal Power Generation Co. together with six other companies including the Kumamoto Electricity Co. Spurred by the establishment of this thermal power generation company, a cooperative relationship was built among the electric power companies and an electricity supply system was created based on the combined use of hydroelectric and thermal power generation.

Chapter 5, "Corporate Governance and Finance in the Railway Industry during the Meiji Period: the Case of the Kyūshū Railway Co.", by Naofumi Nakamura looks into the relation between the structure of corporate governance and financing with a case study of the Kyūshū Railway Co., a large non-*zaibatsu* company of the Meiji period. As for corporate governance, the analysis of the 1899 dispute over management reveals that the confrontation between large and small shareholders had the effect of enhancing the status and significance of the roles of salaried managers. As for corporate finance, the study corroborated that after the managerial dispute the company was compelled to choose capital increase as a means of fund raising, even though that was more costly than issuing corporate bonds. The study concludes that the characteristics of corporate governance in the Meiji period fall within the category of newly industrialized countries.

Chapter 6, "The Foundation Process of the Bank of Fukuoka: the Yasuda Hozensha Co. and the Wartime Bank Merger", by Yurio Mukai is a detailed analysis of the events which led to the establishment of the Bank of Fukuoka under the so-called "one prefecture one bank policy" pursued by the Ministry of Finance and the Bank of Japan during the war. For the Yasuda Hozensha Co. and the Yasuda Bank, their affiliate regional banks such as the Jūshichi Bank played important roles in financing, for instance. Over the merger issue, however, regional banks in Fukuoka prefecture fiercely opposed the Yasuda *zaibatsu*. Eventually, the financial authorities intervened, and thus the Bank of Fukuoka was established. Then, with twists and turns, the Yasuda group finally succeeded in gaining hegemony.

Chapter 7, "The Technological Development of the Shibaura Engineering Works Co. and the Personnel Management of Technical Experts", by Hiroshi Ichihara studies the Shibaura Engineering Works Company's efforts to utilize the human resources of both university graduate engineers and shop floor technicians and to develop their work-related abilities. To promote product development, the knowledge of university graduate engineers mainly assigned to designing and development jobs and that of shop floor technicians mainly assigned to manufacturing jobs had to be combined. The study found that a cooperative relationship was created between these two groups of employees. Furthermore, the personnel system to promote shop floor technicians to higher ranks was maintained at the Shibaura Engineering Works in the prewar period.

Chapter 8, "An Aspect of the Labour Management System at *the State-owned Yawata Steel Works in its Foundation Period: a Study on the Steel Works' Associated*

Hospital", by Noriaki Tokisato is an empirical analysis of the functions and roles of the associated hospital of the Yawata Steel Works.

The hospital was regarded as a welfare facility for employees. Therefore, its costs were paid by the government, and from the viewpoint of workforce preservation, work-related injuries and diseases were treated at the government's expense. The hospital also conducted health checks and physical examinations in order to monitor the physical quality of the workforce. This case indicates that the associated hospital played a part in the Steel Works' labour management by keeping physical conditions of individual workers under surveillance.

Chapter 9, "Salters' Associations in the Meiji Period: the Case of the Mitajiri Saltpan Association", by Akihiro Itō reveals, with a detailed analysis of account books, the actual management conditions of the Mitajiri Saltpan Association which covered the largest area and production volume in Yamaguchi prefecture. In 1878, the Mitajiri Saltpan Association was established by the merger of four associations of salters which had existed since the early modern period. This association consisted mainly of independent salters and aimed not at landowners' control over tenant salters but at independent salters' mutual aid. This association had a distribution organization called "urisabakijo" and a financial function to deal with its members' common property called "hojikin". These functions made possible the salt industry's continuous development during and after the Meiji period.

The significance of this book, as Professor Ogino states in the introduction, is that it is a "study of the region", to produce results from the "regional" point of view and to show how a regional joint research system should function, against the recent research tendency towards "centralization" (p. vi). The special merit of this book is thus the detailed empirical researches on corporate activities at the regional level based on primary sources and related records.

I have one criticism. The contributors share a common understanding of two of the three main themes of the book: (2), the analysis of the national economy from the viewpoint of 'regional development, and (3), the comprehensive analysis of the "regional economy" from the viewpoint of "corporate activities". However, this is not the case with (1), the "structuralization" of the "central" and the "regional". It is true that the notions of "central" and "regional" are themselves problematic. In attempts to define these concepts clearly, however, the abstract term "structuralization", in expressions such as "vertical structuralization from the central", "horizontal structuralization from the regional", and "structuralization of the central and the regional" (p. ii), tends to obscure rather than clarify the notions. Professor Ogino explains that Chaps. 7 and 8 deal with different subjects; however, he suggests that "the comparison between the management of private companies located in Tokyo and that of state-owned companies located in provincial regions is important for a better understanding of the management of companies in the heavy industries in this period" (p. v). The reason why such a comparison might be important requires further clarification. In sum, the merit of this book lies in the detailed empirical studies on the regions. The reviewer hopes that there will be further results coming from the "regions" through the steady efforts of empirical researchers.

Chapter 8
Review of J. R. McNeill, *Something New Under the Sun: An Environmental History of the Twentieth-Century World* (20 Seiki Kankyō Shi. Translated by Masatomo Umitsu and Tsunetoshi Mizoguchi)

Nagoya Daigaku Shuppankai, Nagoya, 2011

Shoko Mizuno

In the past decade or so, 'environment' has been picked up as a theme at many conferences of academic societies including the Socio-Economic History Society. It can be said that attempts have already been made to rewrite the history from the viewpoint of the interaction between humans and the environment, reconsidering the conventional historiography which has dealt only with human societies.

In recent years, there has been a movement to write an overview of the global history involving both humans and the environment beyond the framework of a state. Some of the books have been translated into Japanese, including Arnold (1996) and Hughes (2001). This book by J. McNeil, however, is a representative work on this subject which has been frequently cited by other researches. It is a significant event, therefore, that the translation of this book is published.

The contents of the book, shown below, indicate the aims and characteristics of this work.

1. Prologue: Peculiarities of a Prodigal Century

   Part One: The Music of the Spheres

2. The Lithosphere and Pedosphere: the Crust of the Earth

The review that is translated here, and that originally appeared in *Shakai Keizai Shigaku*, 79(2) (August 2013), pp. 149–151, was of the Japanese-language version of the book. The page numbers have been altered so that they refer to the English original, and some other minor adjustments have been made.

S. Mizuno (✉)
Shimonoseki City University, Shimonoseki, Japan
e-mail: sh2014675@gmail.com

© Socio-Economic History Society, Japan 2015
S. Sugiyama (ed.), *Economic History of Energy and Environment*, Monograph Series of the Socio-Economic History Society, Japan, DOI 10.1007/978-4-431-55507-0_8

  3. The Atmosphere: Urban History
  4. The Atmosphere: Regional and Global History
  5. The Hydrosphere: the History of Water Use and Water Pollution
  6. The Hydrosphere: Depletions, Dams, and Diversions
  7. The Biosphere: Eat and be Eaten
  8. The Biosphere: Forests, Fish, and Invasions

  Part Two: Engines of Change

   9. More People, Bigger Cities
  10. Fuels, Tools, and Economics
  11. Ideas and Politics

Chapter 1 gives a general idea about the magnitude of changes that happened in the twentieth-century society. Part One analyzes segments of the earth's physical environment: the lithosphere, the pedosphere, the atmosphere, the hydrosphere and the biosphere. Human elements are examined in relation to the causes and results of many examples of major environmental changes around the world. Part Two analyzes the social, economic and political tendencies of the twentieth century in order to explain more comprehensively the interaction between the changes in the environment and those in human societies.

This review poses two questions that concern every research attempt, including this work, to depict an overview of the history constituted by the interaction between humans and global environmental changes. The first question is what kind of 'grand history' can be narrated from the viewpoint of the interaction between humans and the environment. The second question is in what way the history has to be described so as to unravel the complex relationship between humans and the environment.

A summary of the basic themes of the book can clarify the historical picture it presents. Only two topics from Part One are touched upon here due to space limitations. The first one concerns pollution. The use of fossil fuels, the cause of air pollution (Chaps. 3 and 4), was accelerated by industrialization, urbanization and motorization. In the twentieth century, air pollution expanded beyond an isolated area as can be seen in the case of acid rain, which had been a regional problem. In addition, although air pollution has become less serious in North America, Western Europe and Japan since the 1970s, it has become a gravely serious problem in those megacities, such as Mexico City, that have grown rapidly since the latter half of the twentieth century. Accordingly, Chap. 4 points out that the air pollution problem is spreading all over the world.

Water pollution (Chap. 5) is similar to air pollution in many ways. It, too, was aggravated and spread by industrialization and urbanization. In the cases like the Rhine river pollution in which a small number of affluent countries like Germany, France and Holland are concerned, international regulations to contain the pollution have been effectively applied since the 1970s. On the other hand, in the cases like the pollution of the Mediterranean Sea in which opposing or struggling countries like Greece, Turkey, Syria, Israel and Egypt are concerned, it is difficult to establish international cooperation.

Another topic is the ecological system altered by humans. In particular, the relationship between humans and microorganisms (Chap. 7) is claimed to have changed rapidly and fundamentally. By the discovery and production of antibiotics, vaccination and immunology, humans have reduced the risk of getting diseases and prompted population growth. At the same time, however, humans have also created environments favorable to the spread of diseases by activities such as the extension of irrigation, accelerated transportation, tropical deforestation, and the development of large cities. It is suggested that the new system humans have created to fend off diseases is only fragile and uncertain.

Similarly, changes in agriculture also had a great impact on the ecological system of the twentieth century. Humans succeeded in making a rapid increase in productivity by means of irrigation (Chap. 6), chemical fertilizer (Chap. 2), mechanization and breed improvement (Chap. 7). The Green Revolution of the 1960s and the 1970s, in particular, boosted food production in Latin America and in Asian countries. However, if the ecological and social consequences of such changes are taken into consideration, the author argues, the fragility of the twentieth-century agriculture will be noted.

In Part Two, it is claimed that the most important and direct cause of the environmental changes of the twentieth century was unprecedented economic growth. Three social factors are elucidated in this regard. The first factor, population increase (Chap. 9), has often been regarded as the primary cause of the environmental changes. It is pointed out, however, that it should not be considered an independent factor and that it was in combination with other factors such as migration and urbanization that the population increase became the driving force for the environmental changes. The second factor is the conversion to the energy system depending on fossil fuels, especially oil, after the 1950s (Chap. 10). The new system developed together with the economic system and technological developments which brought about mass energy consumption, for example, the use of automobiles, industrialization and Fordism. It is maintained, in addition, that the rapid economic integration of the late twentieth century and the further development of the international division of labor prompted environmental changes at the global level. The third factor is the belief in economic growth which spread widely during the twentieth century (Chap. 11). In almost every country, whether it belonged to the capitalist camp or the communist camp, economic growth became an ideology fully integrated into the socio-political system. Consequently, it became difficult to eliminate it, even though the disappearance of ecological buffer zones and the increasing social costs are well recognized.

As shown above, this book demonstrates that though humans have changed the environment around the world in the course of the twentieth century, we still depend on the environment we have altered, and that the relationship between humans and the environment is in a fragile balance. The author claims that the picture of modern history drawn on the premise that the life-sustaining system of the earth has existed only for humans and has been maintained securely is not only incomplete but also misleading (p. 362). If we reconsider the twentieth-century history in terms of the interaction between humans and the environment, the author's claim should be taken seriously.

This book describes the global history of the environment by focusing necessarily on a limited number of cases of environmental changes around the world and examining their causes and consequences. Therefore, the plausibility of the overall argument hinges on the selection of case studies, and the book's attempt to narrate a 'grand history' is in this way limited. However, since it is based on richly empirical case studies, this book presents examples showing critical patterns of environmental changes. By doing so, it succeeds in highlighting the main themes of the environmental history of the twentieth century. The overall view depicted by this approach is, if not novel, persuasive enough as a 'grand history', fairly reflecting previous researches accumulated so far.

As for the next question concerning how the environmental history has to be described, the idea of the environmental history requires us to convert from the traditional historiography, which has focused only on human societies and neglected environmental factors. This book emphasizes time and again the necessity to "integrate history and ecology". That is because the author considers it important to have both the viewpoint of history, which analyzes the intricate mechanism of factors affecting the global environmental changes, and the viewpoint of ecology, which reveals the characteristics and the fragility of the ecological situation specific to the twentieth century.

This book attempts to integrate history and ecology by adopting a two-part structure in order to approach the subject from both directions. While Part One describes "the modern ecological history of the planet" (Preface, p. xxii), in other words the changes in the world's environment from the viewpoint of ecology, Part Two focuses on "the socioeconomic history of humanity" (Preface, p. xxii), in other words the changes in human societies. However, the seven topics examined in Part Two, including the population growth, the economy and the politics, have already been more or less discussed in Part One. The repetition makes the attempt to grasp the whole picture frustrating. The two-part structure of the book may not be functioning effectively for the purpose of unraveling the complex, interwoven relationships between the history of the earth and that of humans.

The integration of history and natural sciences is one of the challenges of environmental history. Therefore, it is important to reflect on problems caused by the differences in viewpoints and approaches of different academic disciplines. This book stimulates thinking about the difference between 'people' in history and 'humans' in ecology. From the ecological point of view, which looks at humans situated in a broad ecosystem, such problems as ethnicity and gender are overlooked. It is not that the author is neglecting the question of what 'people' are. In fact, he repeatedly points out how one environmental problem affects people in various ways.

However, he maintains in a conclusion to his arguments about the biosphere that the defense system against diseases and the changes in agriculture are more important problems than the changes in forests and fishery, because the former problems concern the life and death of billions of people. He goes on to claim that from the viewpoint of mankind, the standpoint of billions of people takes precedence over that of millions, or tens of millions, even if there exist other opinions. Thus the

Dayak tribe in Borneo expelled from the patrimonial estate due to deforestation, and the Californian fishers who lost their jobs in the sardine fishery industry see things differently (p. 265). How to reconcile the difference in the notions of 'people' and 'humans' and bring about a more comprehensive understanding is a crucial problem for the attempt to construct environmental history. This book could have included more in-depth discussion on this matter.

Nevertheless, there is no doubt that this book succeeds in showing how environmental problems have been generated and in giving a view for the world's future, against the criticism that traditional history has not been actively committed to the various issues plaguing modern societies. In this sense, this is more than a general guide for the environmental history. I would like to recommend this book to a wide range of readers beyond the academic discipline.

Finally, a word about the "Preface for the Japanese Version", which surveys the development of the environmental history since the year 2000 when the first edition of this book was published. Among the factors the author has cited in the book as the causes of the environmental changes in the twentieth century, the author emphasizes in this preface the energy system as the most important factor. Examining the changes in the energy system after 2000, he points out the following three phenomena: the gradual development of wind and solar power generation in Europe, the rapid spread of coal use in China and the restoration of nuclear power generation amid the discussion over the reduction of greenhouse gas emissions. In particular, the author repeatedly refers to the fact that the growth in China's carbon dioxide emissions is canceling out the curtailment the rest of the world has achieved. He projects that China's move will be the most crucial factor for the environmental history of the twenty-first century. As for nuclear power generation, although he recognizes the grave impact of the Fukushima nuclear accident of March 2011, he thinks that it is still too early to predict whether the accident will prevent the restoration of nuclear power generation. In any case, finding ways to change the present energy system, which largely depends on fossil fuels, is the most serious challenge for the world of this new century.

References

Arnold, David. 1996. *The problem of nature: Environment, culture and European expansion.* Oxford: Blackwell.

Hughes, J. Donald. 2001. *An environmental history of the world: Humankind's changing role in the community of life.* London: Routledge.

Index

A
Air pollution, 73, 86, 88, 95, 96, 106, 128

B
Birmingham, 90, 92, 97, 99
Brunton, W., 92, 97–99
Buxton, 103

C
Charcoal, 5, 11–25, 32–34
Chestnut sleepers, 38
Coal, 3–5, 20, 22–26, 33, 51, 87–88, 90, 93,
    97–99, 103, 114, 118, 124, 131
Coal smoke, 86–88, 90–89–91, 94, 106
Coal-smoke pollution, 89
Common lands, 4, 6, 7, 11, 14, 22, 26
Commons, 4–7, 9, 22, 26
Cornwall, 88, 101, 102

D
Deforestation, 3–11, 18, 26, 73, 76, 131
Drinkwater, 99
Dugdale, 91, 92

E
Earl of Liverpool, 91
Ecological environment, 69–81
Ecology, 70–72, 74, 75, 77, 81
Ecosystem, 71, 81
Environment, 4, 6, 26, 70, 72–74, 86, 113–115,
    127–131

F
Finlay, K., 91, 92
Firewood, 3–7, 9, 11–26, 32–34
Fuel economy, 98, 99, 103

H
Hasegawa Timber Co., 33
Hiba timber, 53

I
Ina railways, 52
Industrial Revolution, 86–88, 89, 91, 95, 96,
    103, 106

J
Japanese Government Railways (JGR), 33, 35,
    36, 53, 57, 65, 66

K
Kodate Timber Co., 58

L
Littleton, E.J., 102, 103
London, 88, 89, 91, 93, 96–98, 101–103, 105

M
Manchester, 90, 92, 99
Manchester Police Act, 89, 91, 99
Mills, C., 91, 92

© Socio-Economic History Society, Japan 2015
S. Sugiyama (ed.), *Economic History of Energy and Environment*, Monograph Series
of the Socio-Economic History Society, Japan, DOI 10.1007/978-4-431-55507-0

Ministry of Railways, 52
Mitsui, 59
Monteith, H., 91–93, 95

N
Nagano railways, 52
Nagoya Railway Sleeper Limited Partnership
 Co., 53
Natural environment, 66
Nippon Railway Co., 61

O
Okayama railway district, 59
Ominato Timber Co., 53

P
Parkes, J., 92, 95–98, 102, 104, 105
Police Commissioners of Manchester, 89
Pollution, 4
Public nuisance, 86–90, 95, 105, 106
Public pollution, 72
Pulpwood, 33, 34

R
Railway Authority (*Tetsudōin*), 49
Railway nationalization, 37–49
Railway operators, 57
Railways, 31–66
Railway sleepers, 31–66
Reforestation, 5–7, 9, 11

S
Select Committee on Steam Engines and
 Furnaces, 86
Silk reeling, 3–6, 11–14, 16, 18, 20, 22, 25, 26,
 124

Sleeper(s), 32–38, 48–53, 58–61, 64–65
 dealers, 53–58, 64, 65
 quantity of, 59, 66
 shortages, 59, 61–65
Sleeper-producing companies, 53
Sleeper-timber-producing regions, 34
Smoke Nuisance Abatement Act, 85–106
Smoke nuisances, 89–93, 95, 99, 104, 105
Socio-ecological history, 80
South Staffordshire, 88, 102, 103
Steam boilers, 12, 14, 16, 25, 26
Steel sleepers, 61

T
Taylor, M.A., 90–93, 95, 96, 100, 101,
 103–106
Timber, 20, 32–38, 52–54, 58–59, 66
 Class 1, 60
 Class 2, 60
 Class 3, 60
 demand, 64–65
 market, 66
 preservation, 63
 sales, 54
 shortages, 60–61
 supply, 54, 57
Tōbu railways, 52
Tokyo branch sold timber, 57
Tokyo railway management bureau, 59, 65

U
Urbanization, 86, 88, 128, 129

W
Watt, J., 88, 91, 99, 102
Wood, 31–34
Wooden sleepers, 37